NATURAL HISTORY EXCURSIONS IN TENERIFE

Isoplexis canariensis

Dedicated to all our friends in Tenerife
who shared with us
their knowledge and love of their island

Canary Palm

NATURAL HISTORY EXCURSIONS IN TENERIFE
A Guide to the Countryside, Plants and Animals

Myrtle Ashmole
and
Philip Ashmole

KIDSTON MILL PRESS
SCOTLAND

First published in 1989 by
Kidston Mill Press, Peebles,Scotland, EH45 8PH

© 1989 Kidston Mill Press
ISBN 0 9514544 0 4

A CIP catalogue record for this book is available from the British Library

British Library Cataloguing in Publication Data
Ashmole, Myrtle
    1. Natural history excursions in Tenerife.
    I. Title II. Ashmole, Philip
    574.964'9

ISBN 0-9514544-0-4

Line drawings: Hilary Forbes
Maps and diagrams: Myrtle Ashmole
Cover photograph: Philip Ashmole

Printed in Great Britain by Carnmor Print and Design, Preston, Lancs

CONTENTS

MAPS AND DIAGRAMS

DRAWINGS

ACKNOWLEDGMENTS

Although most of this book is written from personal experience and research, had it been left at that it would inevitably have been very incomplete. We were fortunate, therefore, in having the help and encouragement of many people on the island, who generously supplied us with information and suggestions and accompanied us on several trips. We are particularly grateful to Pedro Oromí, Keith Emmerson, Aurelio Martín and Cristóbal Rodríguez, who shared with us much of their knowledge and love for the island and gave us considerable help at the manuscript stage. Juan José Bacallado gave us helpful information about marine life and butterfly distribution; Tómas Cruz also taught us about the marine life; José Luis Martín skilfully led us down lava tubes; Francisco García Talavera showed us a fossil dune; Alberto Brito and Geoff Swinney helped us with the fish section; Marcos Baéz helped us prepare a key to Tenerife dragonflies; Brian Upton put us on the right lines with descriptive geology; Brinsley and Vicki Burbidge and J.H.Dickson made helpful comments on the plant section and Antonio Machado stimulated our thinking with many a long discussion on island biology, and read through the manuscript for us. We include a reading list in the Appendix, but we would like to acknowledge here the extensive use we have made of "WILDFLOWERS OF THE CANARY ISLANDS" by David and Zoë Bramwell, and *"FAUNA MARINA Y TERRESTRE DEL ARCHIPIELAGO CANARIO"* edited by Juan José Bacallado. If there are factual errors in the book, they are our responsibility and not that of any of the many people who have helped us. We would welcome any corrections, which should be sent to Kidston Mill Press, EH45 8PH, Scotland.

Vieraea laevigata

PREFACE

As the title implies, this book is about natural history. However, we have interpreted the words 'natural history' as including more than just plants and animals, and we give a brief account of the volcanic history of Tenerife, its geography, climate and the evolutionary history of its plants and animals. We also include some basic practical information for visitors, which we hope will help with their decisions on where to stay and with their excursions into the countryside.

The Canary Islands are remarkable for their wealth of unique plant and animal life and, but for a stroke of fate, might have played a part in the development of ideas on evolution in the nineteenth century. On 27th December 1831 Charles Darwin set sail with Captain FitzRoy of the "Beagle" for a four year surveying voyage around the world. It is well known that this voyage, and especially his visit to the Galápagos Islands, formed the basis for his ideas on evolution. It is less well known that his first stop was to have been Tenerife. Just as the ship's crew were preparing to land they were brought news that they would have to spend 12 days in quarantine prior to landing and so the captain decided to set sail again. In his diary Darwin writes:- "And we have just left perhaps one of the most interesting places in the world, just at the moment when we were near enough for every object to create, without satisfying, our utmost curiosity." We are more fortunate because we can land without quarantine regulations and explore this fascinating island.

Although the seven Canary Islands are geographically associated with Africa they are politically an integral part of Spain, in much the same way that the Shetland and Orkney Islands are a part of Great Britain. They form two of Spain's provinces; the western group (Tenerife, La Palma, Gomera, Hierro) form the province of Santa Cruz de Tenerife, while the three eastern islands (Gran Canaria, Fuerteventura, Lanzarote), together with five islets, make up the province of Las Palmas. The two provinces together are administered by the Canary regional government *(Gobierno Autónomo)*. Tenerife is the largest of the islands and is now densely populated with about 610,000 people or 297 per square km (768 per square mi).

The variety of altitude, climate and habitats (low semi-desert, forests and high mountains) within the island of Tenerife presents endless opportunities for exploring. However, published information is scattered and can be difficult for the visiting naturalist to find although there are some excellent books in Spanish published locally. Furthermore, the various maps available often omit small roads, while villages and mountains may vary in name and even in position! This guide is written both for visitors to Tenerife and residents, but primarily from the viewpoint of a northwest European who is unlikely to be familiar with even the commonest roadside plants on the island. It is a delicatessen of the natural history of Tenerife - not a complete guide to the plants and animals.

The book is in three sections. The first is a general account of the natural history of the island, including its geography, climate, volcanic history and the origin of its plants and animals. The second section is in guide book form, with practical details about how to get to a wide variety of interesting habitats, and descriptions of what you might see there; these are only brief accounts and are intended to be read in conjuction with the relevant parts of the first section. This is a personal selection but it includes all the main habitat types and should enable the visitor to see a high proportion of the most interesting plants and animals. For instance it includes excursions to places where there is a good chance of seeing the **Blue Chaffinch, White-tailed Laurel Pigeon** and **Bolle's Laurel Pigeon** - three of the four bird species which are found only on the Canary Islands. In several cases we suggest more than one excursion to a given type of habitat, bearing in mind that visitors staying in the southern coastal resorts have access to different areas from those staying in Puerto de la Cruz in the north. We give bus information (in the Appendix) where appropriate, but a car is essential for a few of the places. Several of the places that we discuss are excellent starting-off points for walks, but this is not a walk guide and most places do not necessitate extensive walking.

The third section is for reference and contains annotated lists of the plants and animals mentioned elsewhere in the book, together with some notes on identification and distribution. As far as the plants are concerned, we have concentrated on the more conspicuous ones, but at the same time we have tried to convey some of the excitement that we feel at being able to stand surrounded by plants which occur only here on the islands, and nowhere else in the world. Land vertebrates are treated comprehensively; we include all the breeding birds, many of which are Canary Island specialities, all the mammals, reptiles and amphibians and also the fresh-water fish. In addition, we have provided a list of marine fish found in the waters around Tenerife. Among the invertebrates we include all the butterflies, hawkmoths and dragonflies, but treat the other groups selectively.

HOW TO USE THIS BOOK

This book was written because we felt a need for it, although we were lucky in having plenty of time to learn our way around Tenerife and often local naturalists to help us. We hope that the book will make it easier for people with only a little time to get to some of the special places on the island, and to enjoy them more.

The three sections of the book have quite distinct uses:-

The Natural Environment

Section 1 provides the background. This is the place to look for general information about the island - its geography, natural history and something about its origins. Even if you cannot go on any of the excursions, this section should serve to make your stay on the island more interesting.

Excursions

In Section 2 we describe 24 excursions. They are arranged in natural groups: visits to the lower zone, pine forest, laurel forest and the high mountain zone, plus some extras. If you have only a few days to spend exploring you may wish to choose one from each group, so as to get a feel for the remarkable diversity of the island. The map on page 56 shows where the excursions are. The description of each excursion is preceded by details of how to get there, sometimes with a detailed map. The relevant bus information can be found at the end of the book, just before the index.

You don't have to be an expert naturalist to enjoy these places, and we hope you will not be put off by seeing a fair number of names of plants and animals in the text: some people do want to know these and we have tried hard to give the most useful ones. The "Note on Names and Terms" on page 13 explains why some names are written in different print.

Plants and Animals

Section 3 is designed for those with a special interest in some branch of natural history. If you want to check up on a group of plants or animals, go straight to the relevant part of this section. If you want to know about a particular species you may do best to use the index, but please read the note on the first page of it.

A NOTE ON NAMES AND TERMS

For a popular book it is always difficult to decide what names to use for animals and plants. We have tried to use familiar names wherever possible. Established English names are given priority.

All the birds have English names, and if a species breeds on the island its name is printed in bold type and with initial capitals (eg **Sardinian Warbler**); group names (eg shearwater) are in ordinary type without initial capitals. English names in bold type are also used for reptiles and amphibians, butterflies, moths and some dragonflies. For fish and invertebrate animals that do not have well established English names we use the scientific (Latin) names; following the normal convention these are printed in italic type (eg *Bombus canariensis*).

For plants the situation is more complicated, but there should be no confusion if the following conventions are borne in mind:- English names are used where they are well established, in bold type (eg **Canary Pine**). In two cases we have "invented" English names (**Small-leaved Holly** and **Large-leaved Holly**). Where there are no established English names, but only a single species is being discussed, we refer to plants by their generic names (eg **Daphne, Phyllis**): many of these names will be familiar to gardeners. In a limited number of cases where we need to distinguish between more than one member of a genus, but there are no convenient English names, we use well known Spanish names; these are printed in bold italic type (eg ***Cardón*** for *Euphorbia canariensis* and ***Tabaiba*** for *Euphorbia obtusifolia*). In other cases where several species are involved we give up and simply use the Latin names, in light italic type (e.g. *Bystropogon canariensis* and *Bystropogon origanifolius*). In Section 2, all plant species mentioned in the text are included in the list of notable plants at the end of each excursion, in alphabetical order of their Latin names.

Section 3 provides systematically arranged lists of all the Tenerife plants and animals mentioned in the text, with Latin, English and Spanish names as available; some recent changes in the Latin names are also indicated.

We have gone to considerable lengths to avoid using unfamiliar scientific terms; however a few seem unavoidable and need defining.

"Macaronesian". This is the collective name given by some scientists to the five groups of islands in this part of the Atlantic Ocean - Azores, Madeira, Salvages, Canaries and Cape Verdes. The term sometimes also includes a small part of the NW African coast but for the purposes of this book we use it just for the islands.

"Mediterranean". This word is used to describe the distribution of plants and animals that occur in European and African countries bordering on the Mediterranean Sea.

"Genus". Closely related species of plants or animals may belong to the same genus. For example the two tree heaths on Tenerife are *Erica arborea* and *Erica scoparia*. *Erica* is the "generic" name for both of them.

"Species". Each kind of plant or animal that can consistently be distinguished from its closest relatives has its own name. Thus, with the tree heaths *arborea* and *scoparia* are "specific" names, but these are normally used only in conjunction with the name of the genus or its initial - *Erica arborea* or *E.scoparia*.

"Subspecies". Some species of plants and animals differ slightly when they live in well separated places and scientists often distinguish them as different subspecies. On the whole we avoid discussion of these, although we do include the bird subspecies in Section 3. Thus the Tenerife subspecies of the **Great-spotted Woodpecker** *Dendrocopos major* is *Dendrocopos major canariensis*.

"Endemic species". This is a species that only occurs in a particular place. Thus "Macaronesian endemic species", "Canary endemic species" and "Tenerife endemic species" all describe different limited distributions. In this book, where we use the word "endemic" unqualified, we mean a "Canary endemic". (In a similar way there are also endemic genera and endemic subspecies.)

Geological terms are explained in the relevant part of GEOLOGY, Section 1. The few descriptive botanical words and abbreviations that we use are explained in SELECTED PLANTS, Section 3.

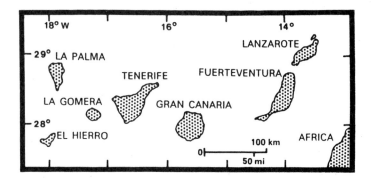

Section 1. THE NATURAL ENVIRONMENT

GEOGRAPHY

There are five groups of islands in the eastern central Atlantic Ocean: Azores, Madeira, Salvages, Canaries and Cape Verdes. Collectively these archipelagos are frequently called Macaronesia (see A NOTE ON NAMES AND TERMS). The Canary Islands lie close to the northwest coast of Africa: Fuerteventura is only 92km (57mi) away. This subtropical archipelago, which is comprised of seven main islands, is over 485km (301mi) from east to west; it lies between 27.5 and 29.5 °N and 13 and 18 .5 °W and the total area of the islands is 7501km² (2896mi²). Tenerife is the largest of the islands and is about 84km (52mi) long, having an area of just over 2057km² (794mi²), about the size of Cornwall.

Tenerife occupies a central position in the Canary Island archipelago, with three islands to the east of it and three to the west; it is 60km (37mi) from Gran Canaria and 27km (17mi) from Gomera.

The island is roughly triangular in shape, and is dominated by Mt.Teide in the centre. Teide is 3718m (12198ft) high (almost three times the height of the highest mountain in Britain - Ben Nevis 1343m (4406ft)). The peak rises from the northern edge of a vast *caldera* - Las Cañadas - which is about 16km (10mi) across, and 75km (47mi) in circumference and at an altitude of 2000m (6500ft) (SEE GEOLOGY). Extending northeast from Las Cañadas is a dorsal ridge of high mountains which are dissected by deep ravines leading towards the sea on either side. The geologically more ancient eastern and western peninsulas of Anaga and Teno are also mountainous; here the ridges and gorges are even more deeply eroded, and in some places there are sea cliffs over 500m (1600ft) high. The southern tip of the 'triangle' is quite different. With a backdrop of mountains to the north, this area is a low coastal semi-desert plain stretching several kilometres inland towards the foot-hills of Las Cañadas and the dorsal ridge.

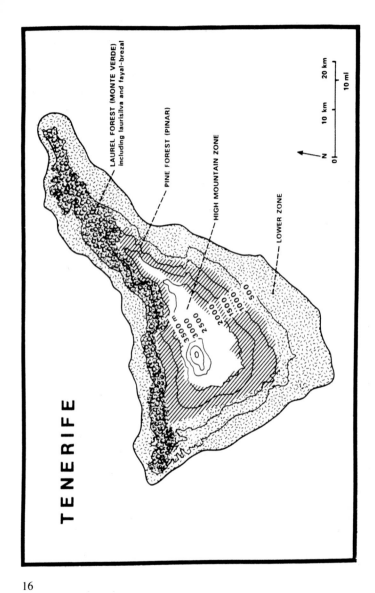

TENERIFE

LAUREL FOREST (MONTE VERDE)
including laurisilva and fayal-brezal

PINE FOREST (PINAR)

HIGH MOUNTAIN ZONE

LOWER ZONE

N

0   10 km   20 km
0   10 ml

GEOLOGY

The western and central Canary Islands, including Gran Canaria and Tenerife, are true oceanic islands, built up from the ocean floor by millions of years of volcanic activity and never connected to the African continent. The eastern islands of Fuerteventura and Lanzarote are of disputed origin; some geologists believe that they are also oceanic islands while others think it likely that they were once joined to Africa. The latter view was given some support by the discovery on Lanzarote about 25 years ago of fossil eggshells, apparently of Ostriches and Elephant-birds. These groups of birds, which are otherwise known from Africa (and Madagascar in the case of the extinct Elephant-birds), became flightless many millions of years ago. However, some scientists now think that the eggshells may belong to more primitive birds, which might still have had the power of flight.

Modern techniques are now being used to determine the age of the rocks on the islands. In many areas, however, the older rocks are inaccessible because they lie deeply buried by successive layers of lava and ash due to subsequent volcanic activity. It seems likely that the eastern islands are oldest, and some Fuerteventura rocks have yielded dates of more than 20 million years ago, very much older than any that have been found on Tenerife. Here the oldest exposed rocks are in the western and eastern peninsulas of Teno and Anaga: these have yielded estimated ages of just over 7 million years and over 5 million years respectively (although there is one tentative date three times that age). In the central part of the island, around Teide, if there are rocks of this age, they lie deeply buried. Volcanic activity in this part of the island has continued into recent times, eruptions being recorded in 1704, 1705, 1706, 1798 and 1909; doubtless there will be more to come. The most recent eruption was that of Chinyero, west of Mt.Teide, whilst the most recent in the archipelago was in the south of La Palma in 1971.

There are conflicting theories about the origin of Las Cañadas in the centre of the island, and rather than get involved in the arguments, we will briefly describe the theory that has received most attention recently. The region was formed by a succession of volcanic eruptions which continued until it reached a height considerably greater than the present level of Las Cañadas. There was then one enormous volcanic explosion which in effect blew the top off the mountain. When this period of activity ceased, the magma (internal molten rock) cooled and contracted and the floor of the crater sank down to its present level: this type of crater is known technically as a *caldera*. Subsequent volcanic activity on the north rim of the *caldera* produced further volcanoes - Teide and Pico Viejo (Chahorra) - and at the same time destroyed the rim. This story is further complicated by the fact that Las Cañadas is really made up of two craters, one extending from the east of Teide National Park as far as Los Roques de García near the south, the other at a considerably lower level from Los Roques to the cliffs along the west of Las Cañadas. One other theory is that there were two large erosion valleys that eventually joined up (at Los Roques), but that subsequent eruptions blocked the exits and partly filled up these valleys, leaving large, inward draining basins.

The word *cañada* refers to the flat, sandy areas which can be seen along the edges of Las Cañadas, below the cliffs (see *LA FORTALEZA* and *ROQUES DE GARCIA*, EXCURSIONS Nos.17 and 19). Areas like this are rare except in recent volcanic landscapes, since running water causing erosion normally carries the eroded materials down gullies towards the sea. A volcanic crater, however, drains inwards and unless the rim of the crater is breached, the products of erosion accumulate on its floor: it is this process that has produced the *cañadas*. In several places one can see how subsequent lava flows from Pico Viejo and Teide have flowed over the *cañadas*, sometimes reaching the crater wall (see *LAS NARICES DEL TEIDE*, EXCURSION No.20).

Almost wherever you are in Tenerife, with the exception of Teno and Anaga, you will see evidence of geologically recent volcanic activity. In the crater at the top of Teide there are fumaroles - holes in the ground out of which steamy sulphurous gases still escape. In many parts of the island there are cinder cones, some being partially vegetated (see *VOLCAN DE GÜIMAR*, EXCURSION No.3) whilst others are clearly of relatively recent origin and still stand out dark and unvegetated. Elsewhere there are large expanses of rough lava, the youngest of which are frequently almost devoid of vegetation. A lava landscape is a "frozen" picture of one stage in a highly dynamic process. Had a particular lava flow solidifed days, or even hours, earlier or later, the extent and form of the lava flow might have been quite different.

For descriptive purposes eruptive volcanic materials can be divided into three main types - gases, pyroclasts and lava. Gases, mainly water vapour, predominate during an eruption, and subsequently small amounts may continue to escape from fumaroles for hundreds of years. Pyroclasts are the bits of fragmented rock and magma (molten rock) that are thrown into the air by these violently exploding gases. They either fall back as solid or semi-solid lumps near the site of the eruption or travel down-wind as fine particles. The various kinds of pyroclasts are named according to size:- ash (<2mm), lapilli (2-64mm) and bombs (>64mm). Scoria (or cinders) are the dark lapilli, while the light coloured (and lighter in weight) are pumice. (In Tenerife the word *picón* is used for lapilli.) Subsequent eruptions, mineralization and erosion all contribute to changes in the pyroclastic deposits, which can become consolidated, or reduced to fine dust over time. Ash sometimes consolidates immediately on falling and forms a deposit known as tuff.

One type of pyroclastic deposit which is common on Tenerife is called an ignimbrite or ash-flow. This results from a spectacular volcanic phenomenon known as a glowing-cloud eruption. In this, an ash-laden column of erupting gas collapses to form a hot, dense, suffocating cloud of material that rushes down the flanks of the volcano like an avalanche, sometimes at speeds of over 100 mph. Such a flow will normally be channelled down any valley that it encounters, and as it slows down and loses its gas it drops its load of particles as a thick and fairly homogeneous carpet.

The third type of volcanic material, the lava, results from molten magma flowing out, or being thrown out, from a crater or fissure and is of three main kinds. "Pahoehoe" lava is relatively smooth on the surface and is usually formed when the magma is very hot and liquid and moves several miles per hour; in some places the surface is wrinkled or ropy giving rise to the local name *lava cordada*. "Aa" lava is jagged and chaotic; this is the commonest kind on the island and is formed from cooler, slower-moving magma. ("Pahoehoe" lava can change into "aa" lava as it loses gas and cools, but the reverse never happens.) The words "aa" and "pahoehoe" are Hawaiian; "pahoehoe" literally means "lava on which one can walk barefoot!" Finally there is block lava which is composed of large chunks of lava which continued to move after they had cooled, because of the pressure of lava that was still being emitted. One feature of areas of "pahoehoe" lava is the presence of lava tubes. These are tunnel-like caves, often of considerable length. A lava tube is formed when a river of rapidly flowing lava creates a channel to flow in. The lava cools and solidifies at the edges and then at the surface, but the molten centre continues to flow until the eruption stops and an empty tube is left behind (see *CUEVA DE SAN MARCOS*, EXCURSION No.21).

Most of the island is built up of successive layers of lava and compressed pyroclastic material. Since the pyroclasts in particular vary in colour and particle size, layers can be seen clearly in many places, for instance along the sides of the road (C824) that runs east along the mountain ridge from Las Cañadas, or along the motorway in the south of the island. Road cuttings also provide a splendid opportunity to study geological dykes which appear as more or less vertical intrusions of rock through the roughly horizontal layers of pyroclasts (see *TENO* BEYOND *BUENAVISTA*, and *ANAGA* BEYOND *EL BAILADERO*, EXCURSIONS Nos.4 and 9). These dykes were formed from molten rock which was forced up from below, through the layers of previously deposited rocks, during episodes of volcanic activity when sub-surface pressures became especially great.

This is no more than a thumbnail sketch of the geology of Tenerife, and we recommend the bilingual book called CANARY VOLCANOES: TENERIFE (see BOOKS, in Appendix) which does far greater justice to the topic.

OCEANOGRAPHY

The Canary archipelago lies in the path of the cool Canary Current, flowing to the south-southwest, roughly parallel to the coast of northwest Africa. Although the strongest part of the current passes east of Tenerife for a large part of the year, it brings cool water to the islands: in summer the sea surface temperature reaches 22-23°C, but in winter it drops to 17-18°C. As a result of these low temperatures the Canaries lack - in spite of their relatively southerly position - the extreme diversity of marine life often found in areas where sea temperatures remain in the twenties throughout the year. Furthermore the waters around Tenerife are very poor in nutrients: although the northeast trade winds induce upwelling along the African coast, enriching the waters between it and the eastern islands, this has little influence on the sea around Tenerife, so that populations of fish and other marine animals are not especially high. Another crucial factor is the way in which the shores of Tenerife plunge steeply down into the sea, so that the island virtually lacks the kind of shallow submarine shelf which is typically so productive of marine life. In this respect Tenerife is among the poorest of the Canary Islands. The area of surrounding continental shelf (the zone less than 200m (650ft) deep) is smaller than that of the island itself. Only La Palma has a relatively smaller shelf. In contrast, Lanzarote (with its islets) has over 3 square km of shallow sea bed for each one of land.

The constancy of the climatic and oceanographic regime results in the different coasts having very distinct characteristics: the north shore is exposed to the trade winds and frequently suffers strong wave action; the southeast shore usually has current and wind flowing southwestwards parallel to the coast; and the southwest shore is very sheltered.

CLIMATE

The Canaries are subtropical oceanic islands, only about 500km (300mi) north of the Tropic of Cancer. Oceanic islands generally have relatively mild winters and cool summers compared with inland areas at the same latitude, but the climate of the Canaries is also influenced by their closeness to the deserts of North Africa. In general, the climate is of "Mediterranean" type, with a very dry summer and fairly low rainfall spread over the rest of the year. The islands lie near the edge of several weather systems but Tenerife is within the influence of the northeast trade winds for most of the year. These trade winds - *los alisios* or *tiempo norte* - acquire humidity as they blow over the cool Atlantic waters and often give rise to an extensive layer of strato-cumulus cloud, especially as the air rises on striking the north side of the main ridge of the island.

When the trade winds are blowing, the climate of the northern part of the island is totally different from that on the south of the main ridge. The whole of the north coast may be under cloud for days on end while the south basks in sunshine. If you are staying in Puerto de la Cruz, a journey up the Orotava valley will almost always bring you into bright sunshine in Las Cañadas; you may find a similarly striking transition as you pass Erjos on the road west from Icod and then south to Santiago del Teide.

The cloud on the north side of the island forms because the cool moist air brought by the trade winds rises as it makes contact with the warm north slopes of the island, leading to condensation (see illustration). The cloud is prevented from rising far, however, by a temperature inversion (warm air above cool air, in the reverse of the normal situation) resulting from the presence of an upper layer of warmer dry air that comes with the high level northwesterly winds. As a result a dramatic "cloud sea" forms over the northern slopes of the island on most days in the year, somewhere between 600m (1970ft) and 1700m (5589ft) and with a typical thickness of 300m

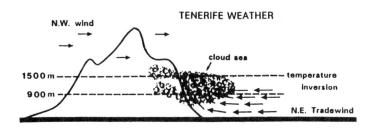

TENERIFE WEATHER

N.W. wind

cloud sea

1500 m — — — — — — — — — — temperature
inversion
900 m — — — — — — — —

N.E. Tradewind

(985ft) in summer and 500m (1640ft) in winter; it tends to be thinnest in the morning and to thicken up throughout the day. This cloud may come as a surprise to those who rely on tourist promotion brochures for their information!

The Anaga peninsula in the east of the island has its summit ridge at about 900m (2950ft), which is not high enough to reach the inversion layer and trap the clouds; there are thus often dense banks of cloud enveloping the upper part of the northern slopes of Anaga, but these continually flow southwards over the ridge, and then evaporate. As a result, the Playa de Las Teresitas, northeast of Santa Cruz, is often just on the border of shade and sun.

The "cloud sea" sometimes clears away completely between November and January, usually when the trade winds give way to cyclonic north Atlantic weather - *tiempo palmero* - with depressions approaching the islands from the northwest and bringing heavy rain or even snow on high ground. Occasionally a tropical wind from the southwest also brings heavy rain. Brief respite from the clouds may also occur at any time of the year when the trade wind is replaced by a gentle southeast wind - straight off the Sahara! This is termed *tiempo sur*, *siroco* or *levante*, and brings with it a fine haze of dust - *calima* - which may almost obscure the sun. The hot air moves in at a fairly high level, leaving a layer below it which is kept cool by the sea: on some of these days Santa Cruz de Tenerife can be cooler than La Laguna, 500m (1600ft) higher up.

Seasonal differences are much less marked on Tenerife than further north in Europe, although there is a definite wet season and temperatures can be quite low in winter. August is the hottest month, with a mean temperature of 25°C in Santa Cruz, near sea level, but only 17°C at the mountain observatory of Izaña at 2400m (7800ft). The coldest month is January with a mean temperature between 17°C and 4°C according to altitude. Frosts and snowfalls may occur from December to March in the mountains, normally only above 1700m (5600ft), and small patches of snow can sometimes be found on Mt.Teide up to the end of May. The difference between the highest and lowest temperatures in any one day is not more than 7°C in most places but often enormously more than this at high levels: in Las Cañadas there can be as much as 25°C difference. One consequence of the low latitude is likely to be noticed by visitors: in summer the days are shorter than further north in Europe (the longest day is 14 hours as against $16^1/2$ in London) but in winter they are relatively long (the shortest day is $10^3/4$ hours as against $7^3/4$ hours).

The average rainfall for a year varies from place to place: it is more than 800mm (31in) in the highest parts of Anaga and lower than 100mm (4in) in the extreme south of the island. Four fifths of the rain falls between

October and March, with July and August the driest months.

The water resources of the island, however, are critically influenced by a phenomenon which is not taken into account by normal rainfall statistics: this is the capture of condensation by the foliage of trees, referred to locally as "horizontal precipitation". As the cloud drifts through the trees the minute water droplets in the air are caught on the cool, shiny surfaces of the leaves and drip off on to the ground below. The trees thus generate their own rainfall. In tests in the cloud zone at Aguamansa, at about 1000m (3280ft) altitude, a rain gauge under trees recorded almost 20 times as much precipitation as a gauge a few metres away in the open. This effect was apparently well known - though not understood - by the indigenous inhabitants of the Canaries (the *guanches*) and one of the famous stories of the Spanish conquest of the islands relates to the accidental betrayal by a local girl to her Spanish soldier lover, of the whereabouts of a providential tree, known as *El Garoé* or "*árbol-fuente*", which apparently supplied water for many of the inhabitants of the island of El Hierro. More prosaically, one can often appreciate the effect from the wet patches on the dry road underneath trees on days when the cloud is down over the forest but it is not actually raining.

Mt. Teide and the cloud sea

ECOLOGICAL ZONES

Throughout this book we consider Tenerife as having four main ecological zones: the lower zone, including both arid scrub and cultivated areas; two types of forest zone at an intermediate level; and the treeless high mountain zone (see diagram, and also map on page 16). This is, of course, a very simplified description of the island's ecology. Each of the zones is described in fairly broad terms in the following pages, while further information can be found in Section 2 of the book, where several places in each zone are discussed in greater detail.

The lower zone, which in most places has very low annual rainfall, extends from sea level up to the lower border of the forest at a height of several hundred metres. It is most extensive along the southeast and southwest sides of the island and in many places it is cut by deep ravines. Most of the towns are in this zone, as well as much of the cultivation.

Above the lower zone the northern and southern parts of the island are quite different. Along the north side between about 500m (1600ft) and 1200m (4000ft) is a zone with almost continuous high humidity and

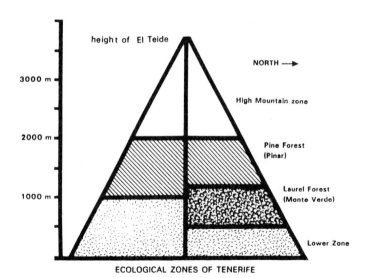

ECOLOGICAL ZONES OF TENERIFE

relatively mild temperatures. This was originally occupied by an evergreen forest in which trees of the laurel family played an important part; now, however, the lower parts are largely destroyed. At higher levels this broad-leaved forest is replaced by pine forests which extend up to about 2000m (6500ft). Along the southeast and southwest sides of the island the greater aridity has prevented the extensive development of laurel forest. Here the vegetation of the lower zone (although in a modified form) extends above 1000m (3300ft) in places to meet the pine forest.

Above the pine forest there is a high mountain zone where the climatic conditions can vary from extreme heat to severe cold within 24 hours. Here the specialized vegetation is dominated by a number of endemic leguminous shrubs, and there is also much exposed ground. The upper slopes of Mt.Teide are distinct from the rest of the high mountain zone, being at a level where there is almost no vegetation.

Between these four main zones there are transitional areas, and in addition to this there are lava fields, cinder cones, and lava tubes or caves at all levels of the island. There is a small amount of fresh water, and additional habitats are provided by the coasts, the intertidal zone and the surrounding sea.

LOWER ZONE

The lower part of the island is characterized by large areas of exposed rock, volcanic cinders and dust, and by more or less scattered plants; many of these plants, especially in the south, are adapted to dry conditions. This zone extends from sea level up to about 500m (1600ft) in the north and considerably higher in the southeast and southwest. In the north most of this land is now urban or cultivated. In the much more arid southeast and southwest, water shortages limit the amount of agriculture and large areas are now only used for grazing (mainly goats). In recent years there has been a great increase in the cultivation of bananas, especially in the southwest, and new plantations are built on raw lava using soil imported from other parts of the island. However, in spite of all this intensive land-use, there are still some areas which are relatively undisturbed. We discuss five broad subdivisions - the coasts, the sandy coastal plains, the Euphorbia communities, the upper transitional areas and the urban and cultivated areas.

Coasts

Tenerife has a very rugged coastline. In some places, epecially in the Anaga and Teno regions, it is virtually inaccessible because of the high cliffs, some of which rise almost vertically from the sea. Much of the rest of the coast can be reached by using well-worn fishermen's paths which make the walking fairly easy. There are numerous coves with small beaches of volcanic sand and many lava fields pushing out into the sea. In a few places ancient coastal sand dunes have been entirely covered by lava flows, but erosion has subsequently exposed the sand at the coast (see *PUNTA GOTERA*, EXCURSION No.22). There are two small islets off the coast of the Anaga peninsula - Roque de Fuera and Roque Adentro - and also one in the north near Garachico; landing on these islets is restricted, but the seabird colonies on them are monitored by local biologists. Most of the Tenerife beaches are pebbly: a few in the north also have dark (basalt) sand whilst the southern beaches are formed from paler, more acidic, rocks (trachite or pumice). One beach of imported pale Sahara sand has been created at Las Teresitas near San Andrés about 8km (5mi) northeast of the capital, Santa Cruz de Tenerife.

Along the coasts in many places one can find a number of salt-tolerant (halophytic) plants which are rarely found further inland. Perhaps the most conspicuous is **Tamarisk** *(Tamarix canariensis)*, a shrub with reddish-brown bark, scale-like leaves and minute pink flowers. At ground level in sandy places (for instance *EL MEDANO* and *MALPAIS DE GUIMAR* - EXCURSIONS Nos.1 and 2) there are two succulent halophytes that also occur in northern Africa. One is **Zygophyllum** *(Z.fontanesii)* which is a low shrub growing about 20cm (8in) high and with grape-like yellow-green leaves. The other is the yellow-flowered **Astydamia** *(A.latifolia)* which belongs to the carrot family but has strikingly fleshy notched leaves. Other typical coastal plants include **Sea Heath** *(Frankenia laevis)* **Crithmum** *(C.maritimum)* and **Limonium** *(L.pectinatum)*.

Sandy coastal plains

A special habitat more typical of the eastern islands of Lanzarote and Fuerteventura consists of sandy coastal plains. On Tenerife it is present only in a few areas in the extreme south (see *EL MEDANO*, EXCURSION No.1). In spite of the rather unprepossesing terrain, this habitat contains an interesting plant community with a number of species not found elsewhere on the island, together with an array of endemic beetles and other inverte-brates associated with them. Two of the shrubs typical of this habitat also occur in North Africa: one is the yellow-flowered composite **Launaea** *(L.arborescens)* which forms a greyish spiny latticework up to about 1m high; the other is **Traganum** *(T.moquinii)*, a species almost entirely re-stricted to sand dunes, which has crowded, fleshy leaves and hairy flowers. Pressure for coastal tourist development is so great on the island that only a very strong conservation policy is likely to save this unique habitat from eventual destruction. One of the more interesting birds of this area, the **Cream-coloured Courser**, is already extremely rare.

Euphorbia communities

In the rest of the lower zone (except of course in the extensive areas of cultivation) one is rarely out of sight of one or more of the endemic euphorbias. These belong to a family of plants - the Euphorbiaceae - which play somewhat the same role in the Old World that the cacti play in the Americas, with many succulent forms thriving in arid conditions. There are a number of kinds of euphorbias on the island, undoubtedly the most spectacular being the tall, cactus-like **Cardón** *(Euphorbia canariensis)* with its vast clumps of succulent, candelabra-like stems; it often grows into impenetrable thickets. Two other species are very widespread and often dominate large areas: these are **Tabaiba** *(E.obtusifolia)* and **Tabaiba Dulce** *(E.balsamifera)* which grows mainly near the coast. Another three species - **Tabaiba Majorera** *(E.atropurpurea)*, **Tabaiba Parida** *(E.aphylla)* and *E.bourgeauana* - are more restricted in distribution. These euphorbias can be distinguished with practice and we describe them in some detail in Section 3.

Although no cacti are native to the Canary Islands, several species of **Prickly Pear** *(Opuntia sp.)* - cacti at one time cultivated extensively as the food plant of the **Cochineal Bug** *(Dactylopius coccus)* - are now found 'wild' in many parts of the lower zone. They flourish in disturbed ground and also invade the wilder areas where they compete for space with the native plants; the seeds are spread both by humans and by various species of birds, including the **Raven**.

In the areas where **Tabaiba** and **Tabaiba Dulce** dominate (for example *MALPAIS DE GUIMAR* - see EXCURSION No.2), the plant and animal communities tend to be relatively simple. One common plant which occurs here is **Kleinia** *(K.neriifolia)*, a thick-stemmed succulent shrub confusingly like a euphorbia but actually a composite. Another common shrub is **Plocama** *(P.pendula)*, the sole member of its genus, and found only in the Canary Islands; it has drooping branches, slender bright green leaves, a strong smell and very tiny flowers. **Plocama** is a successful colonist, its seeds being distributed both by wind and by flood water. Both **Lavandula** and **Asparagus** are quite common. There is also **Periploca** *(P.laevigata)* which is a twining shrub with amazingly long twin seed pods which produce a mass of white fibres when the seeds ripen. Related to **Periploca**, but looking very different, are two species of **Ceropegia** *(C.dichotoma* and *C.fusca)*, which also grow here: they look like sinister fat bits of jointed garden asparagus. In the areas where **Cardón** is the dominant euphorbia (for example in parts of *TENO BEYOND BUENAVISTA*, see EXCURSION No.4), there is a more diverse community with many additional species of plants.

There are very different communities of plants on the cliffs and in the gullies of the lower zone, some of the most spectacular plants being species of **Aeonium**, which are members of the houseleek family. They are extremely variable in growth form. Some are easily recognizable as houseleeks, even if they are as large as the plate-like rosettes - up to 30cm (12in) in diameter - of *Aeonium tabulaeforme* which grows on rock faces (see *TENO BEYOND BUENAVISTA*, EXCURSION No.4). Others are sturdy branched or unbranched woody shrubs, but always with their fleshy leaves in rosettes.

Transitional areas

The upper part of the lower zone was occupied in the past by open woods including the **Phoenician Juniper** *(Juniperus phoenicea)*, a species that also occurs in North Africa and the Mediterranean; scattered individuals can still be found in some areas (see *LADERA DE GUIMAR*, EXCURSION No.6). The **Canary Palm** *(Phoenix canariensis)*, a somewhat more robust relative of the Date Palm, and the **Dragon Tree** *(Dracaena draco)* are also plants of this region of transition between the arid scrub area and the forest, but are now rarely found growing wild. There are a few small groups of **Dragon Trees** growing in inaccessible sites, such as in Barranco del Infierno (see EXCURSION No.8) and Barranco de Masca, and some in a semi-wild situation in Barranco de Igueste (see EXCURSION No.5). Although cultivation has obliterated most of the natural vegetation of the area of transition between the lower zone and the laurel forest, communities of shrubs and small trees surviving in a few places give some idea of what it would have been like in the past (see *TENO* BEYOND *BUENAVISTA*, Point A, EXCURSION No.4).

An artificial transition in the north of the island is formed in the large areas - especially in the Orotava valley - where the introduced **Sweet Chestnut** *(Castanea sativa)* forms woods at a lower level than the native forest.

Urban and cultivated areas

This book is mainly concerned with natural or near-natural habitats. However, urban and agricultural areas do provide opportunities to see several plant and animal species that may be unfamiliar to visitors from further north in Europe. Roadsides and other disturbed areas are good places to become familiar with four of the most conspicuous of the introduced and invasive plants - **Prickly Pear, Agave, Tree Tobacco** and **Castor Oil Plant**. Many towns have the endemic **Dragon Tree** growing in their squares and public gardens, but identifying the **Canary Palm** here is risky because several other species are also grown.

Animals of the lower zone

The invertebrates of the lower zone are of considerable interest to specialists. Some groups are not yet well studied and new species are still being found and described. This is the case, for instance, in the spider family Gnaphosidae, which is well represented in the lower zone; these are nocturnal hunting spiders and you are not likely to find them in daytime except by turning over rocks. Another spider, however, is hard to miss in the lower zone: this is the orb weaver *Cyrtophora citricola*. The female is very large (up to 15 mm long, plus legs), and is black with conspicuous white plates on the abdomen. She makes a large and complex web in **Prickly Pears** and other shrubs: although basically an orb web it is actually shaped rather like an umbrella swathed in masses of tensioning lines. The male is much smaller (about 3mm) but also black; he can sometimes be found in the edge of the female's web. If you look carefully at the web you may see one or more other tiny spiders *(Argyrodes argyrodes)*, with pointed pink and silvery abdomens; these make their living on scraps or small prey caught by the owner of the web.

Visitors with experience of other arid regions may well think of southern Tenerife as suitable country for scorpions. However, scorpions are poor dispersers, and none seem to have reached Tenerife naturally. The only species present on the island *(Centruroides nigrescens)* is native to central and north America, and probably reached Tenerife accidentally in cargo. It is so far restricted to the coastal zone of Santa Cruz; its sting is painful but not lethal. In this connection it is worth noting that although Tenerife is relatively free from dangerous animals, there is a large centipede *(Scolopendra morsitans)* which can inflict a painful, but not fatal, bite; the very poisonous **Black Widow Spider** is rare here but there are a couple of other spiders whose bites may have unpleasant effects. You are unlikely to come across these, but in general large centipedes and spiders are best left to themselves. You are more likely to see, in towns, the occasional very large cockroach, up to 3.5cm (1.4in) long; these are found in tropical and subtropical regions throughout the world and cause major damage to stored food, but are otherwise harmless.

A Canary endemic butterfly, the **Canary Blue**, is widespread in the lower zone, as is the **Bath White**, which may be equally unfamiliar to visitors from northern Europe. Both species of *Danaus* (the **Monarch** and **Plain Tiger**) can sometimes be seen in the lower zone, but they are more typical of gardens and parks such as Parque García Sanábria in Santa Cruz de Tenerife. Of special interest - though of local distribution - is the **African Migrant**, a large greenish white or pale yellow butterfly found in Africa south of the Sahara, which apparently became established in the Canaries only in the mid 1960's, after the introduction of its food plant *Cassia didymobotrya*.

Perhaps the most interesting part of the lower zone for birds is the area of stony plains in the extreme south (see *EL MEDANO*, EXCURSION No.1). It is here that some of the species more typical of the deserts of the eastern Canary islands are to be found. These include the **Stone Curlew, Lesser Short-toed Lark, Trumpeter Finch** and - now exceedingly rare - **Cream-coloured Courser**. Other birds of the lower zone which will not be familiar to visitors from further north in Europe are **Barbary Partridge, Plain Swift, Berthelot's Pipit, Spectacled Warbler, Sardinian Warbler, Spanish Sparrow, Rock Sparrow, Canary** and **Egyptian Vulture** (with exceptional luck!).

One good, but not very attractive, place for bird watching is the area around Los Rodeos Airport near La Laguna; here you are likely to see **Lesser Short-toed Lark** and **Canary**, for instance, and also many other species incuding migrants. The wild **Canary** is not so bright yellow as the caged variety. It is a greyish-yellow bird with streaked black and yellowish underparts and is a Macaronesia endemic which is found only in Madeira, the Azores and five of the seven Canary Islands. In the south of the island near Médano (see EXCURSION No.1) there are one or two reservoirs (sometimes almost empty) where the **Little Ringed Plover** can be found, and a small brackish lagoon where the **Kentish Plover** may be seen and many European waders occur at migration times. Detailed lists of the breeding birds are given in the relevant parts of Section 2.

Wild mammals of the lower zone are not really very exciting; there are only about 10 species and most of these have been introduced. The only native mammals that you might see are bats; only three species have been recorded on Tenerife, but others have been found elsewhere in the archipelago and may well also occur here. In the past bats were frequently seen on the island, but they have declined drastically in recent years, apparently because of the extensive use of insecticides; there is still a chance of seeing them at dusk in parts of Teno and Anaga (see *BARRANCO DE IGUESTE*, EXCURSION No.5). There are two species of shrews - **White-toothed Shrew** and **Pigmy White-toothed Shrew** - with apparently very restricted distributions; they may have been accidentally introduced recently. Five other wild mammals are present in the lower zone, but have probably all been introduced by humans; these are the **House Mouse, Common Rat, Roof Rat, Rabbit** (which is extensively hunted) and **Algerian Hedgehog**.

The endemic **Canary Lizard** is very common. The populations living in different parts of the island differ in several characteristics, especially colour and size, and some biologists recognise three separate subspecies. These lizards, however, are not the giant lizards, which on Tenerife are known only as fossils, although very large species still survive on Hierro and Gran Canaria. By turning over rocks in the lower zone one can find both the **Canary Skink**, which is an iridescent lizard with tiny legs, and the **Canary Gecko**. Geckos have warty skin, large eyes and the ability to turn almost black, or to blanch, quite rapidly. They are active mainly at night, and the special pads on their toes enable them to walk on vertical walls; they can often be seen (or heard) in buildings. Another kind of gecko, the **Turkish Gecko**, has recently become established around the port area of Santa Cruz.

There are only two amphibians on the island and both are widespread in parts of the lower zone. The **Marsh Frog** is only found close to water. The smaller **Stripeless Tree Frog** is often found quite far from water, especially in the banana plantations; this species can frequently be heard "singing" at night and where they are common there is sometimes a continuous background noise for hours on end.

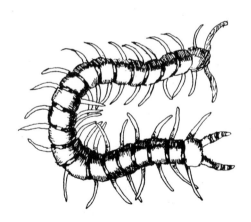

Scolopendra morsitans

LAUREL FOREST - *MONTEVERDE*

The evergreen laurel forest seems to have originally formed a wide continuous band along the north-facing slopes of Tenerife, from a few hundred metres above sea level up to and over the lower parts of the central ridge. It is by no means a uniform habitat and botanists usually distinguish between two main types - *laurisilva* and *fayal-brezal* - although the latter term is used for several quite distinct habitats; *laurisilva* and *fayal-brezal* together are locally known as *monteverde*. The forest has been cut and burned mainly from below, and is now confined to the higher parts of the Anaga peninsula, an even more restricted part of the Teno peninsula, and a few small patches in the central part of the island. The part of this zone that extends from Tejina in the east to Buenavista in the west is now either urban, agricultural, or covered with secondary scrub-forest, so that the true laurel forest has been reduced to less than one tenth of its former area. However, in the best preserved parts of the forest one can still walk in the dim light beneath the dense canopy formed by enormous laurels and other trees of similar form - many of them heavily draped with lichens and other epiphytes - and get a feel for what this part of the island was like in its pristine state (see *ANAGA* BEYOND *EL BAILADERO*, *PICO DEL INGLES* and *MONTE DEL AGUA*, EXCURSIONS Nos. 9, 10 and 11).

Although you may be lucky and visit the laurel forest on a sunny day, the rainfall is higher here than elsewhere on the island; even when it is not raining the forest is frequently shrouded in damp clouds which are whipped over the ridges in a dramatic way by the northerly trade winds, only to evaporate on the southern side. The moisture from the clouds condenses on the leaves, drips to the ground and eventually finds its way down to the underground reserves of water that are so successfully exploited via numerous galleries (horizontal tunnels) and wells throughout the island.

Laurisilva

As explained above, the laurel forest once covered a much greater area than it does today. Although different authorities use somewhat different terminologies, *laurisilva* is the name generally used for the very limited remaining areas of mature forest with many different tree species. There are about 15 species of these broad-leaved evergreen trees on Tenerife, but a few of them are quite rare. Although they belong to 10 different families, the leaves and general growth form of most of them are very similar and this gives the forest a feeling of uniformity; it also makes identifying them quite a problem! Four of the trees belong to the laurel family - hence the name *laurisilva* (*silva* meaning forest).

Nearly all the trees of this forest will be unfamiliar to northern Europeans, since most of them are either confined to the Canary Islands alone (Canary endemics) or to the wider group of Atlantic islands known collectively as Macaronesia. Identifying them all requires hard work. There is an excellent illustrated book on the trees of the Canaries, with an identification key to all the native species; this is by Arnoldo Santos (see APPENDIX). The common species can be learnt fairly easily and being able to identify them adds considerably to the interest of walking through the forest. **Laurus** looks like a Bay (a close relative) and is best recognized by the dull, untoothed leaves with a series of small glands or spots (not just a single pair of large ones) between the midrib and side veins. Another common species is **Prunus** (*P.lusitanica*) which is an evergreen relative of the cherries and plums: it is usually distinguishable by its pinkish leaf stalks and regularly toothed leaves. Two more of the trees are species of holly, but they do not seem much more familiar since the commoner one - **Small-leaved Holly** *(Ilex canariensis)* - lacks prickles (the occasional leaf has a few) and its berries tend to be few and far between; the other - **Large-leaved Holly** *(Ilex perado)* - has very large prickly leaves; in both species the leaves are glossy. One other conspicuous component of the *laurisilva* is a species of tree heath, **Brezo** *(Erica arborea)*; this is easily recognizable as a heath, except for the fact that it is tree sized rather than a low shrub like the heaths of the moorland of northern Europe. In addition to the trees, several characteristic shrubs can be found in the forest. There is also a rich layer of largely endemic herbs, and many ferns, growing especially where breaks in the forest let in a little light.

Fayal-brezal

Fayal-brezal is the name given to the shrub-forest which, unlike *laurisilva*, has relatively few tree species. This type of forest is found along exposed ridges in the laurel forest zone and also in the transitional areas between the laurel forest and the pine forest. The same term - *fayal-brezal* - is used for the extensive areas of cut-over laurel forest that have grown into a secondary shrub-forest, also with very few tree species. It is unfortunate that the same term is used to describe natural climax vegetation and also a habitat resulting from human activity.

The *fayal-brezal* of the transitional areas and the secondary shrub-forest is dominated by two small trees - the laurel-like **Myrica** *(M.faya)* and the tree heath **Brezo** *(Erica arborea)* - thus giving rise to the name. Along the exposed ridges in the forest, a second species of tree heath - **Tejo** *(Erica scoparia)* takes over from **Brezo** (see *ANAGA* BEYOND *EL BAILADERO*, EXCURSION No.9). **Myrica** is of special interest: unlike most of the other broad-leaved trees of the laurel forest it survives in mainland Europe, though only in the Iberian peninsula; in spite of its much greater size, it is a close relative of the fragrant Bog Myrtle (Sweet Gale) of wet heaths in western Europe. **Myrica** is not all that easy to recognise: its leaves are rather similar to those of **Laurus**, but can be distinguished by lack of glands, generally smaller size, and normally being more or less toothed; they are also not unlike those of the **Small-leaved Holly** but are duller and usually narrower. However the edible fruits of **Myrica** (on the female trees only) are a giveaway: small, black, waxy and with a rough - almost faceted - surface. The *fayal-brezal* in the zone of transition between laurel forest and pine forest, and in the undergrowth in the lowest pinewoods, often includes a substantial proportion of **Small-leaved Holly** (see *AGUAMANSA*, EXCURSION No.13).

Myrica faya

Animals of the laurel forest

The *laurisilva* is rich in invertebrates, a very high proportion of them being endemic. Many groups of invertebrates living here are still poorly known, but the ground beetle (Carabidae) fauna has been analyzed and it turns out that almost four-fifths of the forms (species and subspecies) occurring naturally in the *laurisilva* of Tenerife are endemic to the Canaries. Some stocks of these beetles have undergone remarkable evolutionary radiation: for instance, ten species of the genus *Calathus* are found only in the Tenerife laurel forest. The ground beetles are mainly nocturnal hunters, and the best way of finding them is by turning over loose rocks or logs (but replace them carefully afterwards). In areas where *Tejo* is the local tree heath its loose bark forms a wonderful refuge for beetles and other invertebrates. The bugs (Heteroptera and Homoptera) are another group that has been studied in some detail, and they present a picture similar to the ground beetles, in that a very high proportion of species found in the *laurisilva* are endemic. The cultivated areas, on the other hand, have very few endemic bugs but a large number of widespread species.

In contrast to the beetles and bugs, most of the butterflies of the laurel forest are widespread species. Exceptions are the endemic but familiar-looking **Canary Speckled Wood** and the endemic **Canary Blue**. **Red Admirals** can be seen in the forest, but caution is needed since as well as the ordinary species there is the **Indian Red Admiral**, which has one of the oddest distributions of any butterfly (see BUTTERFLIES, Section 3.). Other butterflies typical of the laurel forest are the **Cleopatra** - a close relative of the Brimstone - the **Cardinal** and the **Lulworth Skipper** (the latter two are also found in some parts of the pine forest).

Many of the birds will be familiar to visitors, but a number of species differ in plumage, habits and song from their northern European counterparts and are considered to be distinct subspecies. The brightly marked **Blue Tits**, for example, have a wider range of calls and experienced ornithologists have been known to insist that they have heard Great Tits (which do not occur in the islands); the richly coloured **Robin** is a secretive undergrowth bird and has none of the more familiar boldness. The ornithological highlights are undoubtedly **Bolle's Laurel Pigeon** (also found on Gomera, La Palma and Hierro) and the rare **White-tailed Laurel Pigeon** (also found on Gomera and La Palma). Both of these pigeons feed on the fruits of the laurel forest trees (see comments in *MONTE DEL AGUA*, EXCURSION No.11). Other birds resident in the laurel forest are listed in the relevant parts of Section 2.

The only mammals normally seen in the laurel forest are the **Roof Rat,** feral **Cat, House Mouse** and occasionally the **Algerian Hedgehog**. The **Canary Lizard** may occasionally be seen along tracks and also perhaps the **Canary Skink** if you turn over some rocks.

PINE FOREST - *PINAR*

The second type of forest - the pine forest (or *pinar*) - is found mainly above 800m (2600ft) and up to a maximum of 2200m (7200ft) on the south edge of Las Cañadas. In the past the pine forest extended lower than this in the south, and one can still find outliers down to about 600m (2000ft) in a few places. There is only one native species of pine tree - the **Canary Pine** *(Pinus canariensis)*; curiously, its closest relative is found in the Himalayas, not in Europe or North Africa. Although pines cover a substantial part of Tenerife, it is hard to find areas of "natural" pine forest. This is because felling, replanting and clearing of undergrowth and needles are all part of the island's economy. In fact most of the original pine forest was felled between the time of the Spanish conquest and the middle of this century. Extensive replanting has been carried out in the last 50 years, but until recently up to a fifth of this was of introduced species such as **Monterey Pine** *(Pinus radiata)*. Modern forestry practice - more enlightened from the conservation point of view - is to replant entirely with the native pine. This species is one of the few kinds of pine which are resistant to fire, and is unique in its ability to produce new shoots from the base or trunk after being burnt. Large areas with healthy trees showing severely charred bark can be seen, for example, between Vilaflor and Las Lajas (see EXCURSION No.15). A few groups of very old **Canary Pines** still remain, and it is well worth making the effort to see these magnificent trees (see *MONTAÑA DE JOCO*, EXCURSION No.14). There are two splendid examples by the road side just north of Vilaflor.

The pine forest is not a uniform habitat, in spite of having only one native species of tree. Much of the forest in the north of the island is in the moist cloud zone and the pines grow fairly densely and are often draped in lichens; here the understory is often dominated by **Myrica** *(Myrica faya)* and *Brezo (Erica arborea)* (see *AGUAMANSA* and *EL LAGAR*, EXCURSIONS Nos.13 and 16). At higher altitudes and in the south of the island it is much drier and the forest is more open with only irregular patches of undergrowth and many totally bare patches of rock. **Cistus** *(Cistus symphytifolius)* is the typical undergrowth plant here, whilst in the highest areas of the forest **Adenocarpus** *(A.foliolosus)* and **Chamaecytisus** *(C. proliferus)* dominate (see *LAS LAJAS*, EXCURSION No.15). There are a few places outside the pine forest where **Chamaecytisus** is, in fact, the dominant plant in a shrub-forest community (frequently referred to as *escobonáles* after the Spanish name for this plant, *Escobón*).

Canary Pine

Animals of the pine forest

The invertebrate life is less rich than in the laurel forest, and there are very few species restricted to this zone. The old pines do, however, provide some special microhabitats, and some invertebrates are well adapted to them. For instance, the giant crab spider *Olios canariensis*, which has an extremely flattened body, can be found under the large pads of fleshy bark that clothe the oldest pines; other unusual spiders have been found in the masses of dead pine needles that sometimes accumulate on the lower branches of the pines.

The large and richy coloured **Cardinal** butterfly *(Pandoriana pandora)* can often be seen in the pine forest, which is also the preferred habitat of the endemic **Canary Grayling** and the less conspicuous endemic moth *Macaronesia fortunata*. The hairy caterpillars of the latter, which sport a series of purple tufts down their backs, are enormously abundant on the pines in some years and cause serious defoliation to both **Canary Pines** and **Monterey Pines**. These caterpillars may also be found on bushes of the broom *Retama del Teide (Spartocytisus supranubius)* in some parts of the high mountain zone. Another interesting insect of the pines is the brown-black heteropteran bug *Aradus canariensis*, a very flat species which is adapted to life under the bark of dead trees. In some years there are very high numbers of the large endemic black beetle *Buprestis bertheloti*, although it mainly flies near the tops of trees.

Turning over rocks in the pine forest will expose a number of interesting animals, though few of them are restricted to this zone. One of the most spectacular is the large poisonous centipede *Scolopendra morsitans*. One may also find spiders of the genus *Dysdera* which are easily recognized by their orange legs, brown to purplish carapace and silky grey abdomen. A number of species of this group of spiders have evolved on the archipelago and they have occupied a wide variety of habitats, from the slopes of Mt. Teide to caves near sea level. Two other animals to be found under rocks in the pine forest are the **Canary Skink** and **Canary Gecko**; the latter can also be found hiding in cracks in the bark of the Canary Pine.

It is in the pine forests that you will be able to see the only Canary endemic species of passerine bird that occurs on Tenerife - the **Blue Chaffinch**. On the forested slopes in the north of the island the Tenerife version of the common **Chaffinch** is also present, but any chaffinch you see in the pine forests of the south will be the **Blue Chaffinch**. The Tenerife endemic subspecies of the **Great Spotted Woodpecker** is quite common in the southern pine forest and the easiest way to see both this and the **Blue Chaffinch** is at water drips in picnic sites (see *LAS LAJAS*, EXCURSION No.15). **Sparrowhawks** and **Buzzards** are both frequently seen in the pine forest, and other birds are listed in the relevant parts of Section 2. The **House Mouse** lives in the forest and the feral **Cat** is sometimes seen. The **Canary Lizard** is common in the drier and more open parts of the forest.

HIGH MOUNTAIN ZONE

The high mountain zone starts above the pine forest at a height of about 2000m (6500ft). It includes Teide National Park and the surrounding mountains. The National Park encompasses the vast *caldera* called Las Cañadas, together with the inward-facing cliff walls up to about 500m (1650ft) high and the two volcanoes which have obliterated the northern rim of Las Cañadas - Pico del Teide and Pico Viejo. The *caldera* covers an area of about 12000ha (46.3mi^2), and is 16km (10mi) from east to west and about 10km (6mi) from north to south; its circumference is about 75km (47mi).

People's reactions to the high mountain region differ. It is so alien to anything we see in northern Europe that many people find it stark and unfriendly. On the other hand, many love its austere beauty and return again and again. It is a jagged mountainous region with huge areas of lava and cinders, some of which are ancient and vegetated while others are relatively recent and barren. Although much of the rock is dark grey, there are red, white and yellowish areas; the colours are especially dramatic in early morning or late evening light.

Between the areas of lava, and especially at the edges of the huge *caldera* are the *cañadas*. (The word *cañada* is roughly equivalent to drove-road, and the *cañadas* were used as such during the long period when goats were brought up from the lowlands to graze in the *caldera* each summer.)

The Caldera - Las Cañadas

The *cañadas* are flat 'lakes' of sand composed of granular and powdered volcanic materials. This sand accumulates as the result of erosion by frost, wind and water on the inward-facing cliffs of the *caldera* (see GEOLOGY). After heavy rain or when the snow melts, these areas become temporary lakes, with no outlets: the water gradually seeps down into the ground. During the winter heavy cold air tends to accumulate in the *cañadas* forming frost pockets, where the temperature gets very low at night; in 1911 a temperature of -16.1°C was recorded. It has been suggested that the failure of **Canary Pine** to colonize Las Cañadas may be due to very low night temperatures. The plants and animals that do live in the high mountain zone have many adaptations for dealing with the harsh high altitude environment with its violent fluctuations - both daily and seasonal - between bitter cold and extreme heat, storms of rain or snow and clouds of Saharan dust.

The vegetation of the high mountain zone is dominated by two species of leguminous shrubs - ***Retama del Teide*** *(Spartocytisus supranubius)* and ***Codeso de la Cumbre*** *(Adenocarpus viscosus)* - both Canary endemic species. ***Retama del Teide*** is similar to a large broom bush with white or sometimes pinkish blossom, while ***Codeso de la Cumbre*** is a smaller bush with bright yellow flowers. The other common shrubs of this area are described in Section 2 (see *LA FORTALEZA*, EXCURSION No.17). The best time to see the flowers is in May, June and early July when the colour throughout this high region is quite spectacular.

The proportion of endemic species of plants in the high mountain zone is among the highest for any region in the world. About a dozen species of vascular plant are found almost exclusively in Teide National Park, and a high proportion of the other native plants are either restricted to Tenerife alone or to the Canary Islands as a whole. The walls of the *caldera* are especially interesting (see *LA FORTALEZA*, EXCURSION No.17) and are the refuge for some very rare plants. One intriguing species that we have seen growing near the summit of Guajara - the highest peak on the rim of the *caldera* - is the **Whitebeam** *(Sorbus aria)*, one of the few trees native to both the Canaries and northern Europe (including Britain and Scandinavia). Near the peak of Teide only very few plant species grow; these include the endemic **Teide Violet** *(Viola cheiranthifolia)*, **Gnaphalium** *(Gnaphalium* sp.), the occasional **Argyranthemum** *(A.teneriffae)* and two grasses *(Poa annua* and *Vulpia bromoides)* as well as a few mosses and lichens.

Animals of the high mountain zone

The invertebrate animals in the high moutain zone, like the plants, include a high proportion of endemic species. A few of these are very conspicuous, for instance the **Canary Blue** butterfly, which is to be seen during many months of the year, and whose larvae feed on the numerous leguminous plants. The **Green-striped White** is also common here. The bushes of *Retama del Teide* sometimes seem alive with a shrub-living grasshopper *(Calliptamus plebeius)*, whilst a ground-living grasshopper *(Sphingonotus willemsei)* is sometimes equally evident underfoot; these are both Tenerife endemics, and the first is the mainstay of the diet of the **Kestrel** in this zone. A hairy caterpillar (of the moth *Macaronesia fortunata*) is often abundant on *Retama del Teide* bushes; this is the caterpillar that causes considerable destruction in the pine forests in some years but it seems to cause no lasting damage to *Retama del Teide*. In the spring months it is hard to avoid seeing a large endemic black beetle *Pimelia radula* subsp. *granulata* stumbling across the pathway; it is often to be found, along with **Canary Lizards**, trapped in empty bottles left lying around picnic sites.

Canary Lizard

This subspecies occurs only in the high parts of the island and is very common. A spectacular longhorn beetle *(Hesperophanes roridus)* is also a typical inhabitant of the areas dominated by *Codeso de la Cumbre* and *Retama del Teide*; it is more than 3cm long, grey and square-headed.

Birds that have been recorded breeding in the high mountain zone are **Kestrel, Barbary Partridge, Rock Dove, Plain Swift, Berthelot's Pipit, Spectacled Warbler, Chiffchaff, Great Grey Shrike** and **Raven**, with **Blackbird** and **Blue Tit** around the houses at El Portillo. A few additional native species are sometimes seen but apparently do not breed here; these are **Long-eared Owl, Hoopoe, Robin, Blue Chaffinch** and **Canary**. Although no birds breed on the highest slopes of Teide, a nest of **Berthelot's Pipit** was found at 2500m (8200ft) in 1978. The **Plain Swift** can be seen overhead, but the island population of the **Red Kite** that used to be found in Las Cañadas is now extinct and the **Egyptian Vulture** is no longer seen here.

The **Canary Lizard** is very common in this zone, whilst the **Canary Skink** is found only occasionally. There are no truly native mammals here except for bats, the most likely one at this altitude being the **Canary Long-eared Bat**. However, introduced **Rabbits** are very numerous and are hunted by the islanders for sport. The **Algerian Hedgehog**, the **House Mouse** and **Roof Rat** all occur here, but all these species have reached the island with the help of humans. The most recent and controversial introduction is of **Mouflon**, wild sheep from the Mediterranean (see IMPACT OF HUMANS). Feral **Cats** and **Dogs** both survive in Las Cañadas, probably depending on rabbits for their food. Although many abandoned or lost dogs die, the few that survive and breed have truly wild offspring. They are 'discouraged' by the park wardens and usually keep well away from people, but we have come across pairs clearly associated with rocky lairs.

Lava flows and cinder cones

Whilst the main habitat types on the island are restricted to particular altitudinal zones, volcanic activity is no respecter of height; lava flows and cinder cones occur at all levels. However, because much of the recent activity is in, or only slightly below, the high mountain zone we mention these habitats here. There have been three periods of volcanic activity since the early part of the 18th centrury; from 1704 to 1706, in 1798 and in 1909. The eruption at Chinyero in the west of the island is the most recent, whilst the flow of Las Narices was the result of an eruption high on the western slopes of Pico Viejo in 1798 (see *LAS NARICES*, EXCURSION No.20).

Lava can be incredibly rough and jagged and often with very deep sheltered cracks and semi-caves. In contrast, cinder cones which can be seen in several places on the island are relatively smooth and exposed. Volcán de Fasnia (near the meteorological station of Izaña) is a recent (1705) one but the area around it is now a military firing range and great care should be taken when visiting it, because unexploded shells are found there from time to time. The volcano we include in Section 3 is actually an ancient one in the lower zone (see *VOLCAN DE GUIMAR*, EXCURSION No.3).

When an eruption occurs, hot lava and cinders kill all life in the immediate area and produce a totally sterile environment. The colonization of such areas by plants and invertebrates is a very slow process. The speed depends upon the structure of the lava (or cinders) and on several climatic factors which affect its gradual break-down and the build-up of soil. Plant seeds and spores of mosses, ferns and lichens are carried in by the wind, and a few of these eventually succeed in growing; a varied community of plants is thus gradually built up over a period of tens, hundreds or thousands of years. However, it has now been shown that plants are not necessarily the first colonists in these areas. We have found that recent lava and cinder cones have a surprisingly varied community of invertebrate animals, which are scavengers and not herbivores (see *LAS NARICES*, EXCUR-SION No.20). They depend for their food entirely on airborne insects and debris. Plants come later and are then followed by other invertebrates that depend upon them for their food.

OTHER HABITATS

Caves

Investigations by local scientists in the last few years have shown that the volcanic caves of Tenerife provide living space for an array of fascinating invertebrate animals highly adapted to conditions in these inhospitable places. Both lava tubes and volcanic pits occur on the island, but tubes are much more common. They can be found high on Teide at over 3200m (10500ft) down to sea level at Playa San Marcos (see *CUEVA DE SAN MARCOS*, EXCURSION No.21). These tubes were formed at the time of volcanic eruptions that produced lava flows of the "pahoehoe" type (see GEOLOGY) and are anything from a few metres to several kilometres long; they are usually quite close to the surface. Until recently, Cueva del Viento, near Icod, was thought to be the third longest volcanic cave complex in the world. The recent discovery of more passages increases its total mapped length to 15km (9.3mi) bringing it up to first place, ahead of Kazumura Cave in Hawaii and Leviathan Cave in Kenya. The pits, which are up to 50m (165ft) deep are often cavities left by small volcanic eruptions.

Lava tube

Except in the immediate vicinity of entrance holes, no plants grow in the caves, although the roots of plants that have grown on the upper surface of the lava often penetrate through cracks in the roofs of lava tubes. In spite of the lack of plants, a number of insects, spiders and their relatives live in this environment, among the most interesting being an eyeless beetle *(Eutrichopus martini)*, an eyeless cockroach *(Loboptera subterranea)* and several blind spiders *(Lepthyphantes oromii* and at least two species of *Dysdera)*. Food for these animals either blows in, is carried in by rain water from above or, in some cases, is provided by the plant roots. Bones of the now extinct **Giant Rat** and also the extinct **Giant Lizard** have been found in some of the caves.

Perhaps more surprising than the presence of animals in the volcanic caves is a recent discovery by biologists at La Laguna University. They have found various specialized insects and other invertebrates in the network of tiny crevices that are sometimes present in fairly dry volcanic debris a few metres below ground. Many of the animals normally found only in caves are evidently able to travel through these crevices and so colonize caves that form in new lava flows.

Freshwater

Except on occasions of very heavy rainfall, most of the river beds - *barrancos* - on the island are now dry. Natural pools and a little natural running water can still be found year-round in a few places (see *BARRANCO DEL INFIERNO*, EXCURSION No.8), but most water is intercepted high up and carried in channels and pipes down to the agricultural areas, villages and towns. However, an artificial freshwater habitat is created by these channels, and there are numerous water tanks in the lower part of the island and a few large artificial pools, such as the one near Bajamar (see *PUNTA GOTERA*, EXCURSION No.22). Apart from supporting plants that depend on a moist environment, this habitat is the breeding place for dragonflies and the **Stripeless Tree Frog** and **Marsh Frog**. The pools are also the places to see **Moorhen** and **Little Ringed Plover**. There are a few places, such as Barranco de Igueste (see EXCURSION No.5) where the **European Eel** *(Anguilla anguilla)* can be found; they are the only native freshwater fish (and spend the first and last parts of their life at sea). Two other species of fish - **Guppy** *(Poecilia reticulata)* and **Western Mosquitofish** *(Gambusia affinis)* - have been introduced to help control mosquito larvae, and some pools contain **Common Carp** *(Cyprinus carpio)* and **Goldfish** *(Carassius auratus)* which have been introduced for sport.

Intertidal zone and sea

As discussed earlier (see OCEANOGRAPHY) the waters around Tenerife do not have as high a diversity of marine life as their relatively low latitude might suggest. However, exploring the intertidal zone around the island can be rewarding for the visitor more used to the marine fauna of more northern waters.

The sea is frequently rough and can be dangerous for swimming in several places, but there are some localities with relatively safe access to rock pools at low tide; a snorkel and mask (or diving gear) obviously open up much greater possibilities. Tenerife has no marine nature reserve, although there is a protected area off the north coast of Fuerteventura and one is in the planning stage off the island of El Hierro. Most of the bays have rocky promontories where inshore fish can be seen - in the north try Punta de Teno, Punta del Hidalgo, Playa de San Marcos, Playa de Socorro, Playa del Bollullo, Bajamar and Roque Bermejo; or in the south, Puerto de Santiago, La Caleta, Playa de las Galletas, Playa del Médano and the harbour at El Médano. Several of these places also have rock pools exposed at low tide, but in Section 2 we describe just one of these, which is in the north (see *PUNTA DEL HIDALGO*, EXCURSION No.23). The sea is warmer in the south and actually 1-2 degrees centigrade warmer inshore than the published isotherms would indicate; a few species of algae and invertebrates occur in the south which are typical of more tropical waters.

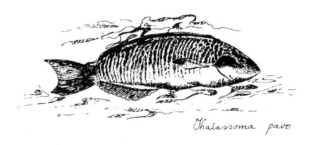

Thalassoma pavo

To many naturalists, the surrounding sea is of interest primarily for the seabirds; since many of these are best observed away from the shore we include details of the ferry journey to the island of La Gomera (see EXCURSION No.24).

Many of the invertebrates and fish that occur in the coastal waters are included in the relevant parts of Section 3. Occasionally whales or dolphins may be seen, or even sea turtles; these are also listed in Section 3.

HISTORY OF THE PLANTS AND ANIMALS

Early stages of colonization

Tenerife and the other western Canary Islands are oceanic in origin; the two eastern islands, however, may have been connected to Africa in the past. Tenerife must have had a violent birth as a volcano emerging from the sea and then being built up to its present great height by a long series of cataclysmic eruptions. At first it would have had no plants or animals. The change from a sterile heap of lava and volcanic ash to the island that we see today, with its complex array of living things, took several million years. This process has been studied by looking at what has happened on more recent volcanic islands such as Surtsey, off Iceland.

Arrival of animals, and of the spores and seeds of plants, begins immediately land is available, but few of these can survive at first because of the austere conditions. Mosses and lichens can start to grow on weathering rock surfaces, while seeds of salt-tolerant coastal plants that drift to the new island in ocean currents have a chance of growing successfully on beaches. Once a few plants are established, their roots help to speed up the weathering process and their dead leaves help in the formation of soil. At this stage other plants have a better chance of becoming established and these create habitats for small animals.

Plants and animals reach a new island by a variety of means, and some are better colonists than others. Among plants, those with seeds that blow away in the wind or can float in salt water are at an advantage. In the case of animals, wings are a help, and so is the ability to travel on floating logs. Less obvious methods are also important: seeds may be transported in the digestive tract of a migrant bird, and some minute animals frequently travel hanging on to bats, birds, or winged insects.

Although it may seem that there is only a ridiculously small chance of a founding group of animals, or of a plant (or its seeds), making the sea crossing to an oceanic island, it is worth remembering that the time available is very long. A quick calculation shows, for instance, that even if only one new stock of plants reached Tenerife and became established in every 10,000 years, there would still have been plenty of time for the modern flora to accumulate: in practice, the better colonists probably arrived relatively soon.

Many colonizing plants and animals, of course, may have been exterminated by later eruptions that covered parts of the island with lava and hot ash. This may explain the absence of land tortoises from Tenerife today. The discovery of a fossil tortoise about 80cm (31in) long in ancient ash deposits near Adeje in southwest Tenerife shows that these animals managed to colonize the island, presumably by floating, or drifting on natural rafts, but they have not survived to the present.

Origin and evolution of the plants

Many of the plants on Tenerife also occur in the Mediterranean-North African region, or are Canary endemics (i.e. found only in the Canaries) but with their closest relatives in that region. Although Tenerife has never been joined to the continents, the fact that both the winds and the ocean currents often flow from northeast to southwest must have helped many North African plants and animals to colonize the island.

A recent check list shows that there are over 1860 plant species growing wild in the Canaries. Many of these have been introduced - intentionally or by mistake - during the long human occupation of the islands. However, the dramatic statistic is that just under 500 species - well over a quarter of all the wild plants - are endemic to the Canary Islands, while almost 60 more are found only on the Canaries and on one or more of the other archipelagos of Macaronesia (the Azores, Madeira, the Salvage Islands and the Cape Verde Islands). This means that wherever you go on Tenerife, you can see plants that grow wild only on these Atlantic Islands and nowhere else in the world.

In order to understand why Tenerife - and the other Canary Islands - have such a special flora we need to go back in time. During the early part of the Miocene period, over 20 million years ago, much of northern Africa was covered by tropical rainforest and savanna, and humid subtropical forests extended into the Mediterranean area. Several species of trees that are now found only in the laurel forests of the Macaronesian islands are known as fossils - or have close fossil relatives - in rock strata of southern Europe. These include three of the laurels occurring in the Canaries (**Apollonias, Ocotea** and **Persea**), the **Small-leaved Holly**, two trees in the family Myrsinaceae (**Pleiomeris** and **Heberdenia**) and one in the olive family (**Picconia**). Although the fossils have been found in southern Europe, it is likely that these trees colonized the Macaronesian islands from parts of Africa adjacent to the eastern Canary Islands, reaching these first, and later the more westerly islands as they were created by volcanic activity. The climate of northwest Africa subsequently changed dramatically with the approach of the Pleistocene ice ages. This led to southward retreat of many plants, but those that had managed to colonize the islands were insulated from the full impact of the changes by the milder oceanic climate and by the high relief of many of the islands, which ensures that a wide range of conditions is available at any time.

The archipelagos thus acted as refuges for some elements of the ancient flora that are now extinct in the Mediterranean and western Africa. As you walk through the damp forests of Anaga or Teno, with their dense canopy of laurels, you are close to stepping back into the subtropical evergreen forests of 20 million years ago.

The story was not finished, of course, when the colonizing plants reached the islands. It was after their arrival that many of the endemic plants of the Canaries acquired the special characteristics that distinguish them from their nearest relatives. In some cases a single stock has split into a number of separate species during evolution on the islands, either on different islands or even within a single one. As a result different islands have fairly distinct floras: Tenerife, for instance, has 108 endemic species not present on any of the other islands. In some plant groups evolution has really gone wild: the succulent genus **Aeonium**, for instance, has 14 species on Tenerife, all endemic to the Canaries and many to Tenerife alone, as well as many more on other islands. This process, in which a single plant or animal stock reaches an environment with few competitors, and then evolves to form a series of species occupying different habitats or ecological niches, is known as adaptive radiation: many of the best examples come from archipelagos.

Origin and evolution of the animals

We can learn a good deal about the process of colonization by looking to see which kinds of animals have managed to reach the islands. Mammals were virtually absent from Tenerife before the arrival of humans: with their high energy requirements most mammals are poor long distance swimmers, and they to do not survive long on rafts. The only ones that definitely reached the island naturally were bats and one rodent. The rodent, which sadly is now extinct, is represented by perhaps the most intriguing of Canaries fossils: named *Canariomys bravoi*, this animal appears to have been essentially a **Giant Rat**, whose ancestors were immigrants from North Africa or Europe. It probably survived until after human colonization of the island, since well preserved bones can still be found in lava tubes.

Freshwater fish and amphibians are also bad at getting to oceanic islands, since they require fresh water for survival. Apart from the **European Eel** *(Anguilla anguilla)*, which spends most of its life in fresh water but breeds in the sea, there are no native freshwater fish on the island. The two species of frogs now present were probably introduced by humans. Reptiles have done a litle better: apart from the tortoise already mentioned, three or four stocks of lizards seem to have reached the islands without human aid. The most spectacular of these was a **Giant Lizard** (see REP-TILES, Section 3) perhaps reaching as much as 1.5m (5ft) in length, which is unfortunately now extinct on Tenerife, though relatives survive on El Hierro and Gran Canaria. A smaller species - the **Canary Lizard** - is still common on Tenerife. Two other types of lizards - the **Canary Skink** and the **Canary Gecko** - are present on Tenerife, and are clearly related to North African and Mediterranean forms. (A second kind of gecko, the **Turkish Gecko**, reached the island recently, evidently with human aid.)

The birds, however, with their great powers of dispersal, have been enormously more successful as colonists than the other groups of vertebrates: the current list of breeding birds on Tenerife amounts to about 56 species. Furthermore, it is very likely that many more bird species once lived on the island: investigations of fossils on Hawaii in the Pacific, and St. Helena and Madeira in the Atlantic, have shown that many birds became extinct - often as a result of human activities - before their presence was recorded by naturalists. Recently a fossil finch, *Carduelis triasi*, has been found on La Palma and some unidentified bird bones on La Gomera.

A number of the birds on the Canaries are given the status of distinct subspecies: this implies that the Canaries populations have evolved substantial differences from their mainland relatives. In a few species divergence has occurred even within the archipelago, so that different islands have different subspecies (e.g. **Blue Tit, Chaffinch, Great Spotted Woodpecker, Kestrel**). In at least four cases evolution of a Canaries bird population has led to the origin of a new species. One of these, the Canary Island Chat *(Saxicola dacotiae)* is not found on Tenerife. The other three, however, are interesting members of the local bird community.

First, the **Blue Chaffinch**, with distinct forms on Tenerife and Gran Canaria, is clearly related to the ordinary **Chaffinch** of Europe, which is also represented on the island, but it does not interbreed with the latter and must thus be considered a separate species. Since only one chaffinch species is present on the continent, it seems likely that the Canaries have been colonized twice by chaffinches: the first arrivals - the ancestors of the **Blue Chaffinch** - increased in size and evolved long, strong bills as they adapted to feeding on the seeds of the **Canary Pine**. A later group of colonists remained separate, living mainly in the laurel forest and mixed woods of the northern parts of the islands.

The second bird species endemic to the Canaries is the **White-tailed Laurel Pigeon**. The ancestors of this pigeon probably inhabited the Mediterranean laurel forests many million years ago, and were responsible for bringing seeds of the trees to the islands in their guts: they would thus have eventually created the habitat in which their descendants survived when the

49

laurel forests on the continent disappeared. The third endemic bird species is another laurel forest pigeon, **Bolle's Laurel Pigeon**, which may have a similar history, but in this case a closely related bird lives on Madeira and some people think that the two forms are subspecies of a single species.

Among the invertebrate animals the situation is similar to that in the vertebrates: in the more mobile groups (as in the birds) many different species have colonized the islands, but in general they have changed rather little after getting there. In contrast, less mobile animals tend to be represented on the islands by only a few evolutionary lines, since colonization events are rare, but once there they have often undergone striking evolutionary changes and in some cases given rise to arrays of closely related endemic species.

The first type is exemplified by the butterflies, with about 25 species on the island. Between 3 and 7 of these (according to the authority) are endemic species, but most of the remainder are widespread in the Mediterranean area or further south in Africa. Among the most interesting are two spectacular butterflies in the genus *Danaus*, the New World **Monarch** and the rather similar Old World **Plain Tiger**. It appears that these insects have both reached the islands without the help of humans, but from opposite sides of the Atlantic Ocean! It is clearly relevant that the **Monarch** is the most impressive of all butterfly migrants, some individuals flying from Canada to Mexico in the autumn.

The other situation, involving less mobile animals, is well shown by the land molluscs: the snails and slugs. Many of the snails on the Canaries belong to two endemic genera, whose ancestors perhaps arrived in mud on the feet of migrant birds, or on floating logs. In one of these genera *(Hemicycla)* no less than 88 species have been described from the islands (although the real number of valid species is probably much lower). Evidently the poor dispersal ability of these animals - though it makes it hard for them to reach the islands - promotes the formation of distinct species on the different islands, and even in different parts of a single island.

These examples may give some idea of the complex development of the plant and animal communities of Tenerife, up to the arrival of the first human explorers. The time since then - negligible in evolutionary terms - has seen drastic ecological changes which are considered next.

IMPACT OF HUMANS

In this book, we attempt to draw attention to the areas of Tenerife that are biologically interesting. There is no evading the fact, however, that human influence has been and continues to be devastatingly destructive to the native habitats. The first settlers arrived on the Canary Islands at least 2000 years ago, and perhaps much earlier: they gave rise eventually to the *guanches,* the natives found on the islands by the western European conquerors at the start of the fifteenth century. The *guanches* were almost certainly of North African (berber) stock, and although they must have arrived by sea they apparently retained no sea-faring tradition; contact between populations on the different islands seems to have been very slight.

Although the *guanche* early settlers probably cultivated barley and a few other crops in relatively small areas, they were primarily pastoral people, herding sheep and goats. The goat - that four-legged ecological disaster of much of the world - probably had an especially heavy impact on vegetation that had evolved in the absence of large grazing animals. It is tempting to wonder what the arid southern part of Tenerife would have been like before the arrival of this animal, and the westernmost part of the Teno peninsula perhaps gives us the best clue.

However, it was probably not until after the arrival of the Spanish conquerors that the landscape of the forested zones began to change dramatically. The native pine and laurel forests, which once covered a large proportion of the island, were cut for timber, firewood, charcoal, pitch and tannins and they are now reduced to a fraction of their previous area. Human influence in the pine forests is still pervasive. The trees are systematically harvested, and some of the finest ancient pines on the north side of the island east of Aguamansa were cut down in 1985 to make a fire break. Fallen needles are collected in most areas every few years, and more frequently along roadsides as a fire precaution. The needles are raked together and then taken down the mountains in lorries for cattle bedding and mulch. This removal of the fallen needles (and also the leaf litter in the laurel forest) destroys the habitat of the invertebrate animals that live on the forest floor.

Extensive sugar cane plantations were established in the sixteenth century and this crop required huge amounts of timber for fuel for the sugar extraction process. Since then, with the increase in population up to well over half a million, there has been a steady increase in the impact of agriculture on the native vegetation. Potatoes, grapevines and **Prickly Pear** (grown for the **Cochineal Bug** which lives on it, as well as for its fruit), replaced sugar cane. These were followed by tomatoes, bananas and other fruit and yet more recently by large acreages of greenhouses with their

51

cucumbers, peppers, more tomatoes and cut flowers. Vineyards continue to flourish in many parts of the island although abandoned mountain terraces have grown back to 'wild' forest of **Sweet Chestnut** or have been re-afforested, sometimes with non-native species of pines and eucalyptus. Several species of **Agave** were introduced for fodder. In some places, especially in the remote fertile valleys of Anaga, 'weekend' agriculture is extensively practised; patches of potatoes and other vegetables flourish here, as they probably have for hundreds of years, in close association with the laurel forest.

Of these crop plants, the **Prickly Pear** stands out as one that has got out of hand and is now spreading throughout much of the arid zone and competing with the native flora. Among other invasive introduced plants are the **Agave, Bramble** (apart from the native Canary one), **Castor Oil Plant** and **Tree Tobacco**. All are familiar roadside plants in the island, whilst some hillsides far from roads have impenetrable thickets of **Bramble**, **Prickly Pear** and **Agave**. Although some species of intentionally or unintentionally introduced plants are more or less confined to disturbed areas many have had adverse affects on the native flora.

Such effects are of special concern in Las Cañadas, with its special community of endemic plants. A recent study recorded about 30 species of flowering plant as successful invaders of Las Cañadas, half of them being grasses; another 50 or so alien species have been recorded there but are not established. Some of these invaders may have originally occurred naturally lower down on the island, while others have been introduced from other parts of the world. All of them, however, have probably been helped by humans in their invasion of Las Cañadas: of the successful species, at least half are thought to have arrived as a result of goat-herding before 1950, a quarter probably arrived as a result of tourism since that time, while the remainder may have come in either way. The species brought by goat-herding probably mostly sprouted from seeds that were in goat droppings or travelled on the feet or coats of the animals, but goatherds are also known to have spread seeds along the tracks to improve the grazing.

Goats and sheep are now excluded from the Teide National Park, but the seasonal influx of herdsmen and their flocks to take advantage of the summer grazing was stopped only about 40 years ago. Although the vegetation shows signs of recovering from this annual attack, the effect of the **Rabbits** (also not native to the islands) cannot be overlooked and the introduction of about 12 **Mouflon** (Mediterranean wild sheep) in 1971 has added to the problems. There are now over 100 of these sheep; not only do they range over areas where there are exceedingly rare plants but they also distribute plant seeds. We feel that it is a pity that a non-native mammal should be introduced into a national park merely for the sport of hunters. Although we are assured that they are no longer encouraged, they can still be seen there. The effect of Rabbits is now being studied, partly by keeping them out of small fenced plots.

Rats have been introduced (presumably accidentally), as in so many other islands all over the world. These may be having a detrimental effect on certain tree species in the laurel forest because the young shoots of certain species are selectively eaten. The presence of rats is probably also one of the reasons why nearly all the colonies of breeding sea birds are confined to offshore islets and cliffs where they are safe from these predators. The other predators on sea birds have, in the past, been the islanders themselves; eggs were taken and the fat young **Cory's Shearwaters** were collected for their oil and for food. This is now forbidden by law (although some poaching continues).

Another example of human predation is that on the large limpet *Patella candei*, which is still present on the two eastern islands but seems now to be absent from Tenerife and the other western islands; its shells, however, are found in shell-heaps left by the early inhabitants on these islands. This case makes one wonder about the possible effect of the current intense harvesting of octopuses. It seems likely, however, that the limpet populations were vulnerable because the animal lived right at the top of the intertidal zone; the octopus, in contrast, lives lower down the shore, and is relatively safe from human predation along many steep parts of the coast.

With the introduction of agricultural and ornamental plants, many associated insects arrived on the island. A large number of these are pest species, so insecticides are used for their control, with the inevitable detrimental affect on native insect species. The periodic influx of locusts is also combatted by large scale spraying. Bats - once common on the island - are now rare, probably mainly because of contamination of their insect prey by insecticides. The vicinity of Santa Cruz de Tenerife is often cited as the place where a new introduction is first seen; this is the case for the scorpion *Centruroides nigrescens* and the **Turkish Gecko**. It was presumably via the port that such common urban pests as rats, mice, some species of cockroach as well as many of the agricultural insect pests, were introduced.

Human activity has also altered the distribution of water on the islands. In addition to the pipes, canals and water tanks mentioned earlier (see

ECOLOGICAL ZONES - Freshwater) there are many very deep wells and over 1,000 horizontal tunnels *(galerías)* with a total length of over 1,600km (996mi) dug into the mountains to intercept underground water. As this vast natural resource is gradually depleted, the wells are dug deeper and many *galerías* are abandoned.

In some places, the landscape is changing rapidly. Whole hillsides are cut away and the soil and fine volcanic cinders from them are transported to the banana plantations which are often created on raw lava fields where almost nothing grew before. However, the most dramatic change on the island in recent years is the steady march of urbanization. The population is now about 610,000 and the birth-rate in the Canaries is the highest in Spain. Much of the lower part of the north of the island is densely populated; recently the main expansion has been in the extreme south. The development here is mainly in response to the huge increase in demand for tourist accommodation. There is no doubt that fragile ecosystems are threatened by this development, especially in the region of Los Cristianos and El Médano. What to us may be an interesting and rare habitat worthy of preservation, to others is only an empty desert area useful for nothing but more tourist developments.

CONSERVATION

We tend to think of concern over conservation issues as a modern phenomenon, but developing ecological problems are more obvious on islands than on continents and in oceans. As early as the beginning of the sixteenth century the local authorities on the Canaries had become aware of the dangers of over-exploitation of their natural resources, and were involved in strenuous but largely ineffective efforts to provide protection for the forests. The *Sociedades Económicas de los Amigos del País*, established on several of the islands in the late eighteenth century, were well aware of the threat to the economy posed by forest destruction. The La Laguna branch offered a prize in 1784 for the best report on the present state of the forests, the causes of their ruin, and possible means for their restoration. Substantial reafforestation programmes, however, were started only in the 1930s; by 1978 13,000 hectares (32,000 acres) had been planted on Tenerife, using mainly **Canary Pines** but also some **Monterey Pines**. As a result of recent changes in attitudes to conservation **Monterey Pines** are no longer planted, and the pines previously planted in Las Cañadas and in parts of the Teno peninsula - naturally treeless areas - have been eliminated. The reafforestation programme - which has led to reestablishment of native pines over large tracts of previously denuded land - has been cited as a striking example of a positive conservation effort.

ICONA (Instituto Nacional para la Conservacíon de la Naturaleza) is the official Spanish organization with responsibility for Teide National Park. Until recently it was also responsible for most of the forest areas but now these are administered by the regional government. Teide National Park was set up in 1954, with the principal aim of the maintenance of the natural ecosystems. The native plants and most birds and animals are protected here by law, and the master plan for the park includes the policy of eliminating exotic plants in the park. However the anomaly still remains concerning the deliberately introduced **Mouflon**. Outside the park, protection laws cover all birds of prey and sea bird colonies; this has come too late for the **Red Kite**, which is now extinct on the islands. Hunting of **Mouflon**, **Rabbits** and game birds is a popular sport and is normally restricted to two days a week in the hunting season. Certain trees on the island are protected - **Dragon Tree, Canary Palm** and both species of **Juniperus**. Furthermore, some parts of the remaining laurel forest are protected from grazing and felling. However, there are many additional vulnerable areas on the island that receive no protection from the developers and speculators. Local conservationists, including groups such as *Asociacíon Tinerfeña de Amigos de la Naturaleza*, have become active in recent years; they have made vigorous attempts, for instance, to prevent the building of unnecessary roads in areas with natural vegetation, and have campaigned for the preservation of places with special ecological significance. In spite of these efforts, however, important sites are still being spoiled, perhaps because not many people realize their long-term value. In 1984, the Tenerife local government commissioned a study which catalogued and discussed 26 areas which should be protected with urgency, if the unique flora and fauna of this island are to be preserved. Whether this produces results remains to be seen.

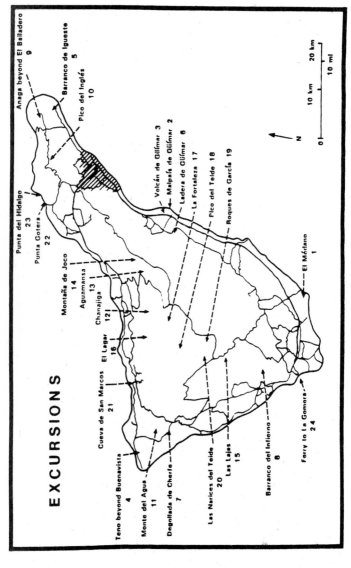

EXCURSIONS

Teno beyond Buenavista 4
Monte del Agua 11
Degollada de Cherfe 7
Cueva de San Marcos 21
El Lagar 16
Chanajiga 12 1
Aguamansa 13
Montaña de Joco 14
Punta Gotera 22
Punta del Hidalgo 23
Anaga beyond El Bailadero 9
Barranco de Igueste 5
Pico del Inglés 10
Volcán de Güímar 3
Malpaís de Güímar 2
Ladera de Güímar 6
La Fortaleza 17
Pico del Teide 18
Roques de García 19
El Médano 1
Ferry to La Gomera 24
Barranco del Infierno 8
Las Lajas 15
Las Narices del Teide 20

N

0 10 km 20 km
10 ml

Section 2. EXCURSIONS

In this section we describe a series of places that are interesting from the natural history point of view. We have tried to include examples of the main ecological habitats that are easily accessible for visitors who are either staying in the north or in the south of the island; we do not, however, include visits to any agricultural or urban areas. We assume the reader has a good road map of the island - preferably the *Cabildo* map (see APPENDIX). However in several cases we give considerable information about how to get to the places, since the maps are not sufficiently detailed.

This is not a walk guide and although a fair amount of effort is needed to get to some of the places, others can be reached by bus or car. Where a visit can be readily combined with a more extended walk we have referred to LANDSCAPES OF TENERIFE by Noël Rochford for further details. Where the places we have chosen are associated with maintained tracks there are usually small yellow diamond shaped signs saying *"sendero turístico"* (tourist path) which will reassure you that you are on the right path.

The discussion of the biology of some of the places is fairly brief and we strongly recommend reading the relevant parts of Section 1, and also descriptions of other places in the same zone, before making a visit. The flora of Tenerife is very rich, and we have made no attempt to mention all the plants that occur in each of these places. We emphasize trees and shrubs, and also the succulents of the lower zone and other conspicuous endemic plants. Some of these are mentioned in the text whilst others are only in the lists at the end of the place descriptions. For conservation reasons, we have not included specific localities of very rare plants. There are over 1280 wild vascular plants on Tenerife (according to FLORA OF MACARONESIA, eds. A.Hansen and P.Sunding); this figure does not include subspecies and hybrids. We refer to only 200, and these are all listed in Section 3, where we also provide brief, non-technical descriptions. Most of the endemic species and a few non-endemic ones are fully described in WILD FLOWERS OF THE CANARY ISLANDS by David and Zoë Bramwell. The bird lists are intended to be complete as far as Tenerife breeding birds are concerned, although we have omitted locations for one or two endangered species; we have not included migrants. The breeding distributions of the birds have been mapped by Aurelio Martín in *ATLAS DE LAS AVES NIDIFICANTES EN LA ISLA DE TENERIFE*. The butterfly lists are also intended to be comprehensive, although keen observers may well be able to expand them in some cases.

We have tried to avoid using too many full scientific names in the accounts of the excursions; where only a generic name of a plant is given, the full scientific name can be found in the relevant plant list; all the plant names are also given in Section 3. The plant and animal name "rules" that we have followed are described earlier in the book in A NOTE ON NAMES AND TERMS.

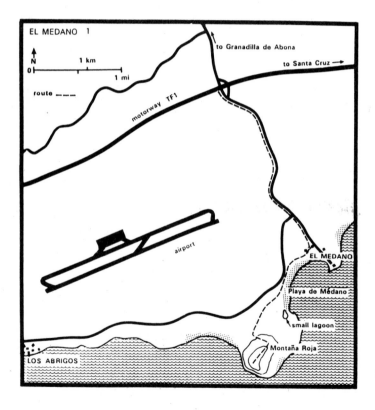

LOWER ZONE (Excursions 1 - 8)

1. El Médano

Habitat description

Coastal site in southern arid part of the lower zone, with a stony plain, wind-blown volcanic sand, and an isolated volcanic cinder cone - Montaña Roja.

Special points of interest

This is the only significant area of the island with wind-blown sand - almost amounting to dunes - and with associated plants which also occur in similar habitats in the eastern islands and in the Sahara. The recent construction of a new tourist resort here makes it a less exciting area to visit from the naturalist's point of view but, together with the nearby arid areas, it is probably still one of the better places to see **Trumpeter Finch, Lesser Short-toed Lark** and **Stone Curlew** and, if you are fantastically lucky, **Cream-coloured Courser**. There is a small lagoon which is one of the few places where migratory waders can be seen.

HOW TO GET THERE. *EL MEDANO* is in the extreme south of the island, close to the airport. **Bus**: See Appendix. **Car**: Take the first motorway exit east of the airport and drive 3.5km (2mi) south to *EL MEDANO*. At the western end of this small town is the *HOTEL PLAYA SUR*. Drive along the track past this hotel towards the isolated volcanic cinder cone - *MONTAÑA ROJA*. There are several drivable tracks in the dune area and it is easy to park.

PRACTICAL DETAILS. Easy wandering but be prepared for heat and wind, and wear stout shoes if you intend to climb *MONTAÑA ROJA*.

The excursion

The area south of the airport between El Médano in the east and Los Abrigos in the west is a habitat unique for Tenerife and of a type much commoner in Lanzarote and Fuerteventura. Unfortunately much of it has been spoilt by bulldozing, sand collection and the construction of a new tourist resort. However the area is still of considerable biological interest and local conservation groups continue to try to ensure that some of it is preserved for posterity.

Close to the shore in sandy areas you may find a number of typical salt tolerant plants. Three of these are the familiar **Sea Knotgrass** (*Polygonum maritimum*), **Sea Spurge** (*Euphorbia paralias*) and **Sea-heath** (*Frankenia laevis*) which grow close to the shore in northern Europe. There are small groves of **Tamarisk**, a plant which is found in southern Europe and in Africa. You will also find three succulent shrublets from the northern African flora; these are the yellowish fleshy **Zygophyllum**, the carnation relative **Polycarpea** with succulent silvery leaves and **Traganum** with tightly crowded fleshy leaves. Where it is rocky along the shore you will find the endemic yellow-flowered **Lotus** *(L.sessilifolius)*.

The flat stony and sandy areas to the north of Montaña Roja, and further along the coast to the west, have a sparse - almost Saharan - vegetation which can look somewhat unimpressive and dry for much of the year; spring and early summer are the best times to visit. One of the dominant plants here is **Launaea**, a dense greyish green small shrub with spiny stems and small yellow dandelion-like flowers.

A small pale yellow and black solitary wasp *Bembix flavescens* lives in holes in the sand here; it can sometimes be seen flying low over the sand, and then landing and burrowing. You might also see on the ground the large predatory bug (2cm long) *Reduvius personatus*, a Mediterranean species (which bites!). If you can find a slightly damp place you will have a chance to find a mole cricket, *Gryllotalpa africana*, a fascinating insect which lives underground and has its front legs specially adapted for digging. Another insect living underground here among plant roots is the nocturnal carabid beetle *Scarites buparius* which is large (over 3cm) and glossy black.

If you walk towards Montaña Roja you will come across a small area of dunes of wind-blown volcanic sand; they are not very impressive dunes but they are the only example of this type of habitat on Tenerife. Montaña Roja can be climbed from this dune area by following the track which starts low on the slopes and climbs gently to the left. This joins with a track which comes up from the eastern side. Several endemic shrubs grow on the slopes. These include the drooping **Plocama** as well as the composite **Allagopappus** which is a small shrub with narrow, sticky leaves and clusters of small flowers. *Tabaiba Dulce* grows here and if you are observant you may find the rare endemic **Kickxia**, a figwort relative forming compact clumps with dramatic yellow spurred flowers. In addition to these endemics there are unfortunately a few of the invasive introduced **Prickly Pear**. Near the top of Montaña Roja, which incidentally does not have a crater, the vegetation is much the same, but you may well notice clumps of **Ceropegia**, with naked stems and reddish-brown flowers.

This southernmost part of the island is a good place to watch out for **Trumpeter Finch, Lesser Short-toed Lark, Stone Curlew** and **Hoopoe**; there have been occasional sightings here of the **Cream-coloured Courser** even in recent years. There are also the **Rabbit, Canary Lizard**, and

Canary Gecko. The latter can be found by turning over a few substantial rocks; but remember to put the rocks back in position afterwards.

The small brackish lagoon near the beach to the northeast of Montaña Roja was one of the few places on the island where you were likely to see migratory waders, and also the resident **Kentish Plover**, although this is now quite rare in Tenerife. However, recent infilling of the lagoon threatens to destroy an important habitat for these birds. The **Little Ringed Plover** breeds in some freshwater pools slightly further inland.

SOME NOTABLE PLANTS. *Allagopappus dichotomus, Astydamia latifolia, Atriplex glauca, Ceropegia fusca, Cyperus capitatus, Euphorbia balsamifera* (**Tabaiba Dulce**), *E. paralias* (**Sea Spurge**), *Frankenia laevis* (**Sea-heath**), *Kickxia sagittata, Launaea arborescens, Limonium pectinatum, Lotus sessilifolius, Ononis serrata, Opuntia* sp. (**Prickly Pear**), *Polycarpaea nivea, Polygonum maritimum* (**Sea Knotgrass**), *Reseda scoparia, Schizogyne sericea, Tamarix canariensis* (**Tamarisk**), *Traganum moquinii, Zygophyllum fontanesii*.

BIRDS YOU MIGHT SEE. **Cory's Shearwater** (over the sea), **Kestrel, Barbary Partridge, Cream-coloured Courser** (very rare), **Stone Curlew, Little Ringed Plover, Kentish Plover, Herring Gull, Rock Dove, Turtle Dove, Long-eared Owl, Plain Swift, Hoopoe, Lesser Short-toed Lark, Berthelot's Pipit, Spectacled Warbler, Great Grey Shrike, Spanish Sparrow, Linnet, Trumpeter Finch.**

BUTTERFLIES YOU MIGHT SEE. **African Migrant** (rare), **Bath White, Canary Blue** (?), **Clouded Yellow** (?), **Monarch, Painted Lady, Plain Tiger, Small White, Small Copper** (rare).

Canary Gecko

2. *Malpaís de Güímar* (Güímar Badlands)

Habitat description

Coastal lava field of relatively recent date, in which the dominant plants are euphorbias. In addition there are some sandy patches with salt tolerant plants.

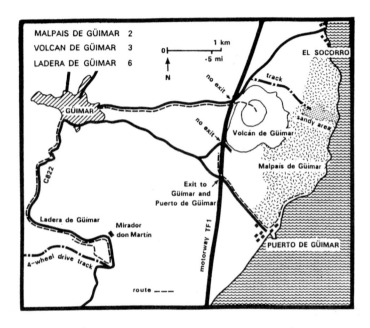

HOW TO GET THERE. *MALPAIS DE GÜIMAR* is halfway down the eastern coast of the island, just south of *CANDELARIA* . **Bus**: See Appendix. **Car**: At *PUERTO DE GÜIMAR*, which is only 1.5km (1mi) off the motorway, turn left along the coast and park at the end of a rough unpaved road.

PRACTICAL DETAILS. Be prepared for heat. The going is very rough if you wander from the coastal fisherman's path, and you will need stout shoes, or preferably boots. Allow at least 30 minutes for a quick visit and considerably more if you plan to wander.

The excursion

The Malpaís de Güímar is one of the very few areas where the plant and animal communities of fairly recent (in geological terms) raw lava at coastal level has been left relatively undisturbed. We hope that all efforts will be made to protect it from the advances of property developers; this area is one of the gems of the island and should be recognised and respected as such. That it is loved by some islanders is evident from the notice you will see, one side of which says *"Que la paz y el espíritu de este jardín vayan contigo."* (May the peace and spirit of this garden go with you.)

The *malpaís* is a mass of jagged "aa" lava (see GEOLOGY, Section 1) between the sea and the base of Volcán de Güímar. Just to the north of Puerto de Güímar a coastal track leads past a few underground dwellings, which are still in use by fishermen and their families. A short way beyond these there are some sandy patches with two of the characteristic salt-tolerant plants, **Zygophyllum** and **Astydamia**.

If you are well shod you can walk inland and across some of the very rough lava. This is an austere but fascinating area, particularly attractive in the low sunlight of early evening. The plant community fairly close to the coast is dominated by *Tabaiba Dulce*, which forms large dome-shaped clumps, often with extensive patches of bare black rock in between them. As you go further inland and uphill you will see more of two other euphorbias, *Cardón* in large clumps and *Tabaiba* singly as erect small shrubs. All the euphorbias have a milky white sap which drips out if you break them. It is very unpleasant and the sap from *Tabaiba* is really dangerous if you get it in the eyes. Sap from *Tabaiba* and *Cardón* has been used as a narcotic for catching fish, and that from *Tabaiba Dulce* to make a form of chewing gum. You may well see where stems have been cut so that the sap can be collected. The alien cactus - **Prickly Pear** - does not seem to have invaded this area yet, as it has so many other parts of the island, and we hope that it can be kept out in the future.

In this habitat, dominated by species of **Euphorbia**, there are several endemic beetles and other invertebrates which are adapted to feeding on these plants, in spite of their poisonous latex. One such animal is the **Spurge Hawkmoth**, whose spectacular striped caterpillars can be found on *Tabaiba*. Inside the dead trunks of the *Cardón* one can sometimes find a particularly interesting insect, the endemic termite, *Bifiditermes rogeriae*: it is a member of a genus which is otherwise restricted to the tropical zones of the Old World. Presumably its ancestors colonized the Canaries at a time when the genus was also present in northwest Africa. Also in the dead trunks of *Cardón* you may see larvae belonging to the endemic longhorn beetle, *Lepromoris gibba*, the adults of which are large and grey but not often seen.

In contrast, a variety of spiders are very visible here and seem to build webs on every available plant.

Judging from the number of droppings, the **Rabbit** manages to live on, or at least visit, the *malpaís*, and the **Canary Lizard** is common. Several species of birds occur here, one of the more interesting being the **Long-eared Owl** which has been recorded nesting on the ground inside large clumps of ***Cardón***. The **Trumpeter Finch** can be found here, and you may see the **Hoopoe** and **Barbary Partridge**; however we have been here on a windy day in mid-summer and not seen a single bird.

SOME NOTABLE PLANTS. *Astydamia latifolia, Campylanthus salsoloides, Ceropegia fusca, Euphorbia balsamifera* (**Tabaiba Dulce**), *E. canariensis* (**Cardón**), *E.obtusifoli*a *(Tabaiba)*, *Lavandula* sp., *Opuntia* sp. (**Prickly Pear**), *Periploca laevigata, Plocama pendula, Schizogyne sericea, Zygophyllum fontanesii.*

BIRDS YOU MIGHT SEE. **Cory's Shearwater** (over the sea), **Kestrel, Barbary Partridge, Herring Gull, Rock Dove, Long-eared Owl, Plain Swift, Hoopoe, Berthelot's Pipit, Spectacled Warbler, Chiffchaff, Great Grey Shrike, Linnet, Trumpeter Finch.**

BUTTERFLIES YOU MIGHT SEE. **Bath White, Canary Blue, Clouded Yellow (rare), Monarch, Painted Lady, Small Copper** (rare), **Small White**.

Ceropegia fusca

3. *Volcán de Güímar* (Güímar Volcano)

Habitat description

Volcanic cinder cone, a fine crater and sparse vegetation adapted to arid conditions.

Special points of interest

Volcanic crater and large volcanic bombs.

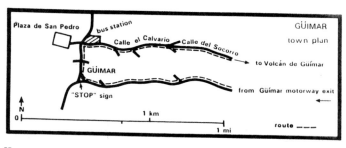

HOW TO GET THERE. *VOLCAN DE GÜIMAR* is halfway down the eastern coast of the island, just south of *CANDELARIA*. See map on page 62. **Bus:** See Appendix. **Car:** Although this cinder cone is close to the motorway, there is no exit associated with it. You could park your car at the *GÜIMAR* exit and walk back to the second fly-over bridge before this exit, or try to find your way by car via unsignposted tracks from the exit to the fly-over. However, we think that the easiest way is to leave the motorway at the *GÜIMAR* exit and drive west into the town and then east again in order to pass over the motorway. Leave *GÜIMAR* by the route shown in the sketch map above; this road does cross the motorway, although this is not indicated on the *Cabildo* map. You can park just the other side of the bridge.

PRACTICAL DETAILS. The track is very steep and much of it is on loose cinders. Stout walking shoes are recommended. Be prepared for heat, and wind at the top. You will need about 20min to climb up the volcano.

The excursion

Volcán de Güímar is a relatively undisturbed volcanic cinder cone, standing isolated in a flat area of the southeastern coast. It has a perfect circular vegetated crater at the top from which there are excellent views up and down the coast: it contains very attractive stands of native shrubs growing in a relatively undisturbed state - although goat grazing must be having an impact on the vegetation. Together with the Malpaís de Güímar (see EXCURSION No.2) it comprises an area that local conservationists are working hard to protect. The volcano can be climbed by following a steep

65

goat track up past a small water tank and on up to the top. At first there is some poisonous **Tree Tobacco,** an introduced weed, but as you walk up you will see entirely native (and mostly endemic) plants, including two euphorbias - *Tabaiba* and *Cardón*, while a third species - *Tabaiba Dulce* - grows at the top. As you walk around the rim of the volcano you will see how *Tabaiba Dulce* grows in very exposed places, while *Tabaiba* appears to prefer more sheltered conditions. The only plant that can survive the extreme conditions on the highest part of the rim is a large member of the dock family - **Rumex.** Lying on the rim and slopes of the crater are some good examples of volcanic bombs (see GEOLOGY, Section 1).

Among numerous invertebrates in this area, one that you are likely to notice is the endemic grasshopper *Oedipoda canariensis*, which is characterized by the pale bluish colour of the hind wings; it is typical of bare rocky or sandy areas exposed to the full sun. The birds you are likely to see are much the same as for the *malpaís* below (see *MALPAIS DE GÜIMAR*, EXCURSION No.2). As elsewhere in the lower zone, you may see the **Canary Lizard, Canary Gecko** and **Rabbit.**

SOME NOTABLE PLANTS. *Allagopappus dichotomus, Asparagus* sp., *Euphorbia balsamifera* (**Tabaiba Dulce**)*, E.canariensis* (**Cardón**)*, E.obtusifolia* (**Tabaiba**)*, Kleinia neriifolia, Launaea arborescens, Nicotiana glauca* (**Tree Tobacco**)*, Periploca laevigata, Plocama pendula, Rumex lunaria.*

BIRDS YOU MIGHT SEE. **Kestrel, Barbary Partridge, Herring Gull, Rock Dove, Long-eared Owl, Plain Swift, Hoopoe, Berthelot's Pipit, Spectacled Warbler, Chiffchaff, Great Grey Shrike, Canary, Linnet, Trumpeter Finch.**

BUTTERFLIES YOU MIGHT SEE. **Bath White, Canary Blue, Clouded Yellow** (rare), **Painted Lady, Small Copper** (rare), **Small White.**

Tabaiba Dulce

4. *Teno* beyond *Buenavista*

Habitat description

Steep north facing slopes, cliffs and gullies near sea level in a geologically ancient massif.

Special points of interest

Four localities extraordinarily rich in endemic plant species, dominated by euphorbias. This is a very popular area for botanists - more than 300 plant species are recorded from the Teno peninsula - and it goes without saying that you should "take nothing but photos and leave nothing but footprints".

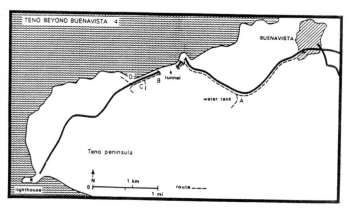

HOW TO GET THERE. The sites we describe are on the northern coast of the *TENO* peninsula - at the western end of the island. **Bus:** See Appendix. **Car:** As you enter *BUENAVISTA* from the east, take the left turning at the cross roads, signed *"TENO"*. For Point A stop after 2.5km (1.5mi) where there are two very large cylindrical water tanks on your left, as the road approaches the cliffs. For Point B continue past the lookout point of *EL FRAILE DE TENO*, where you go under a rock arch; go through the road tunnel (you can walk through this without a torch), and stop at the other end. Point C is 0.6km beyond the tunnel end where there is writing on the cliff face saying *"COTO DE CAZA TF 10.024"* (meaning private shooting). Point D is 0.26km further along the road - just beyond the main cliff and where you first see a continuous slope rising from the sea on your right and on upwards to your left. There is a culvert under the road and a sign in blue on a cement block saying *"COTO DE CAZA"*.

PRACTICAL DETAILS. Sturdy shoes are advisable for scrambling up the gullies. It is best to leave a day, or at least half a day for a trip to *TENO*, especially if you plan to continue on to the lighthouse. Each excursion from the road could take anything from a few minutes to an hour or so, depending on your interest and how far you go. The rocks are often loose and treacherous.

The excursion

There are two flat coastal areas on the north side of the Teno peninsula - one around Buenavista and the other towards the lighthouse on the extreme western point of the island. Between these the mountains come close to the coast and it is in this region that we describe four different places to visit.

Point A. There is a cobbled track here which leads behind the water tanks near the south of the road and up the right-hand gully to the area above the cultivated zone. Three euphorbias - **Tabaiba, Tabaiba Dulce** and **Cardón** – grow here and there are a few **Prickly Pear**. The slopes and gully are particularly rich in shrubs, most of which are endemic. Among the interesting ones is a large **Convolvulus**, with dramatic sprays of white flowers at the tips of the branches: it is widely grown in gardens on the island. You will also see the straggly shrub **Lavatera**, which is an endemic relative of the hibiscus, with delicate pink flowers. Other plants that can be found here are an endemic arum **Dracunculus**, and a **Teucrium**, a fascinating pink-flowered sage relative with several different leaf forms on a single plant; there is also a small endemic St.John's Wort - **Hypericum** *(H.reflexum)*. The euphorbias contain poisonous latex but in spite of this their leaves are the food for caterpillars of the **Spurge Hawkmoth** and in spring almost leafless plants are a common sight.

Higher up in the gully is a fourth euphorbia, *E.bourgeauana* and also **Bramble** and the cane **Arundo**, making use of the somewhat less arid conditions. In this area you may see some plants which are typical of the transitional area between the lower zone and the more humid laurel forest - a zone which has been almost entirely taken over by agriculture in most of the island. A species very characteristic of this situation is **Jasminum**, with glossy green pinnate leaves, elongate yellow flowers and black-brown oval fruit. The **Olive** tree can be found here and also the **Pistacia** tree, with rowan-like leaves; the latter is a member of the cashew and mango family, and resins from some of its close relatives in the Mediterranean have formed the basis of lacquers and a kind of chewing gum since ancient times. On some rock faces near the head of the gully are beautiful growths of lichens.

SOME NOTABLE PLANTS. *Arundo donax, Convolvulus floridus, Dracunculus canariensis, Euphorbia balsamifera (**Tabaiba Dulce**), E.bourgeauana, E. canariensis (**Cardón**), E.obtusifolia (**Tabaiba**), Hypericum reflexum, Jasminum odoratissimum,*

Justicia hyssopifolia, Kleinia neriifolia, Lavatera acerifolia, Olea europaea ssp.cerasiformis **(Olive)**, *Opuntia* sp. **(Prickly Pear)**, *Periploca laevigata, Pistacia atlantica, Rhamnus crenulata, Sideritis* sp., *Teucrium heterophyllum.*

Point B. The tunnel opens out through a magnificent geological dyke, like a partly broken-down gigantic wall cutting across the landscape (see GEOLOGY, Section 1). Growing on the cliffs close to the road is the yellow flowered **Vieraea**, a member of a genus which includes just this one species which is restricted in distribution to this area of Tenerife (it gives its name to a local journal of natural history). Growing alongside it are a species of **Sonchus**, and the plate-like rosettes of one of the most spectacular species of **Aeonium**; you may also see the bushy oval-leaved dock - **Rumex**. Of special interest in this area are the species of **Argyranthemum** (white-petalled chrysanthemum relatives). In the past there were two distinct species in this area, *A.frutescens* on the coastal slopes and *A.coronopifolium* on the wet vertical cliffs. However, material excavated during the construction of the tunnel in 1965 led to the formation of scree slopes nearby, and the two species came in contact and hybridized: at first the hybrids were restricted to the scree slopes, but they are now occurring more widely and *A.coronopifolium* can hardly be found in its pure form.

SOME NOTABLE PLANTS. *Aeonium tabulaeforme, Argyranthemum coronopifolium/ frutescens, Rumex lunaria, Sonchus* sp., *Vieraea laevigata.*

Point C. There are about five rough steps cut in the rock on the left here which give access to another gully much steeper than that at point A and with very different vegetation. A fifth species of euphorbia, ***Tabaiba Parida***, dominates the cliffs with its clusters of naked fleshy stems. There are fewer other shrubs here than at A, but you may see **Ceropegia**, which is somewhat similar to ***Tabaiba Parida*** but with larger stems in less dense clumps. On the slopes you will see the tall stems of **Scilla**, a genus of bulbs well known to visitors to the Mediterranean; and in crevices you may find clumps of a local species of the small succulent **Monanthes**.

SOME NOTABLE PLANTS. *Asparagus* sp., *Ceropegia dichotoma, Cheirolophus canariensis, Echium* sp., *Euphorbia aphylla (**Tabaiba Parida**), E.canariensis (**Cardón**), Lavandula* sp., *Monanthes silensis, Rumex lunaria, Scilla latifolia, Sideritis* sp., *Vieraea laevigata.*

Point D. The gully up above the road here, unlike the gully at C, is well populated with ***Cardón***, and there is also the euphorbia-like succulent composite **Kleinia**. Close by the road you may notice the large shrub **Withania** whose fruits betray its relationship to the nightshades and Tomato. The slope below the road here is a very rich area, with ***Tabaiba Dulce*** dominant in some parts. There are also, however, several patches of

Cardón. This spiny succulent forms impenetrable thickets and hence plants that grow in association with it gain protection from the grazing of goats. Some of the thickets here include **Rubia** - a prickly scrambling shrub - **Justicia, Asparagus, Launaea** and **Lavandula**. Two more of the plants that often grow in the *Cardón* clumps are members of genera endemic to the Canaries: **Neochamaelea** is a small, densely hairy shrub with very narrow leaves and yellow flowers; and **Schizogyne** is a composite with greyish, linear leaves and small yellow flowers. **Schizogyne** is a halophyte (salt tolerant), but is here growing up to almost 100m (300ft) above the sea, along with a halophytic umbellifer **Crithmum** whose light seeds are dispersed in the sea and can easily blow up slopes with onshore winds. Another halophyte, **Mesembryanthemum**, forms bright green and red mats along the roadside, its glandular leaves covered with shiny swollen crystalline hairs. This species and a close relative were extensively cultivated in the island for the soda industry in the eighteenth century, and they were introduced for the purpose. In times of famine, this plant was used to make a substitute for *gofio* (the local flour made from toasted grains). Another interesting plant on this slope is the endemic crucifer **Parolinia** forming compact clumps covered with pink flowers: its closest relatives are in East Africa.

SOME NOTABLE PLANTS. *Asparagus* sp., *Campylanthus salsoloides, Ceropegia dichotoma, Crithmum maritimum, Echium* sp., *Euphorbia aphylla* (**Tabaiba Parida**), *E.balsamifera* (**Tabaiba Dulce**), *E.canariensis* (**Cardón**), *Justicia hyssopifolia, Kleinia neriifolia, Launaea arborescens, Lavandula* sp., *Mesembryanthemum crystallinum, Neochamaelea pulverulenta, Parolinia intermedia, Plocama pendula, Rubia fruticosa, Schizogyne sericea, Scilla latifolia, Todaroa aurea, Withania aristata.*

If you have time it is worth making a trip to the end of the road by the lighthouse, where there is an outstanding view along the cliffs of Los Gigantes on the south side of the peninsula. It is very windy and exposed here. There are magnificent patches of *Cardón* and the **Canary Lizards** living on the rocks near the sea seem to grow as large as anywhere on the island. This coast is one of the few places in the island where you might also see the **Rock Sparrow**. Out to sea you may see **Cory's Shearwater**, the **Little Shearwater** and the Macaronesian subspecies of the **Herring Gull**. Several species of fish are common in the pools; these include such typical rock pool fish as gobies, blennies, puffers, triggerfish and wrasses. Many more inshore fish can be seen, of course, with snorkel and mask.

As this book was being written it became clear that even this remote part of the island will become developed. The road had already been upgraded and a small harbour built near the lighthouse.

BIRDS YOU MIGHT SEE. **Cory's Shearwater** (over the sea), **Little Shearwater** (over the sea), **Buzzard, Kestrel, Barbary Partridge, Herring Gull, Rock Dove, Long-eared Owl, Plain Swift, Berthelot's Pipit, Grey Wagtail, Blackbird, Spectacled Warbler, Sardinian Warbler, Blackcap, Chiffchaff, Blue Tit, Raven, Spanish Sparrow, Rock Sparrow, Canary, Greenfinch, Linnet.**

BUTTERFLIES YOU MIGHT SEE. **African Migrant, Bath White, Canary Blue, Canary Speckled Wood, Clouded Yellow, Large White, Monarch, Painted Lady, Red Admiral, Small Copper, Small White.**

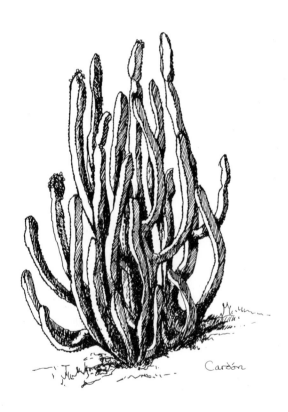

Cardón

71

5. *Barranco de Igueste* (Valley of Igueste)

Habitat description
Valley with dry country scrub vegetation modified by felling, grazing and small-scale cultivation.

Special points of interest
Easily accessible **Dragon Trees** in a semi-wild state.

How to get there. *IGUESTE* is on the south of the *ANAGA* peninsula at the extreme east of the island. **Bus:** See Appendix. **Car:** The road going eastward from *SANTA CRUZ* and through *SAN ANDRES* finishes at a group of small villages, now usually known collectively by the name of one of them - *IGUESTE DE SAN ANDRES*. The road runs about 0.5km inland, crosses the *barranco* and runs back along the other side before ending. At the top of this loop, in the village called *LOMO DE LA CRUZ*, there is now a narrow road along the bottom of the valley.

Practical details. Easy walking, so no special footwear needed unless you intend to explore up the valleys. Once you leave the main road at *LOMO DE LA CRUZ* you need about 40 minutes for a leisurely walk to the first group of **Dragon Trees** and back.

The excursion
Barranco de Igueste, as it approaches sea level, forms a broad valley with steep hill slopes and cliffs on either side. It is reached from Lomo de la Cruz where you can either walk along the new road into the valley or follow the yellow sign, pointing up some steps, saying *"LAS CASILLAS"*. If you climb and keep climbing up paths and steps for about 2 minutes you will reach the edge of the village. Here the way forks. Along the lower path there is a **Canary Palm** with another sign nailed on it saying *"LAS CASILLAS"*. This track joins the new road and three minutes walking brings you to a low red-tiled building and the Barranco de la Sombra de Igueste comes in view on your left. If you turn up this *barranco*, you will see **Dragon Trees** on the slopes to your left: a further walk up this side valley for about 10 minutes will bring you in sight of at least 15 growing more or less wild (the truly wild ones are in very inaccessible parts of the island). The lower part of this valley is partially cultivated, and although it is wilder higher up there is evidence of old terraces in places. Two of the large succulent endemic euphorbias grow here - ***Cardón*** and ***Tabaiba*** - as well as the composite **Kleinia** which is confusingly like a euphorbia. In the same area as the **Dragon Trees** you will be able to see fine examples of **Lavatera**, a hibiscus relative. As you come opposite the first group of **Dragon Trees** you cross a small stream coming in from your right. On the west facing slope of the hillside on your right, a third euphorbia - ***Tabaiba Dulce*** - is very abundant: usually it is

found rather closer to the sea. The stream bed is one of the few on the island that has pools even in mid-summer and is hence a likely place for the **European Eel**, water beetles and the **Grey Wagtail**. It is also a good place for dragonflies: you are likely to see the broad-bodied *Crocothemis erythraea* (which also occurs in southern Europe) and the more slender but also brilliant *Trithemis arteriosa* (which is more typical of the African tropics); these and other species that occur here can be identified by referring to the key and descriptions in Section 3. This valley is one of the only places on Tenerife where there are still bats, which can be seen flying at dusk; the most likely species to see is **Savi's Pipistrelle**. **Barbary Partridges** are quite common here, their strident calls often echoing across the valley. Sometimes they don't fly until one is nearly on top of them, when they rise with a startling whirring noise and fly low over the ground for a short distance.

There is another attractive group of **Dragon Trees** further up the main valley. To reach this, go back to the road by the red-tiled building and walk up it for about 30 minutes (ignore the sign to the right saying *"LAS CASILLAS"*), until you come to another valley with about 15 of these trees on the hillside on your left; there are also a few **Canary Palms** in this area. As you walk along the main track of this valley you will find that the most conspicuous plants are **Asparagus**, the cane **Arundo** and **Bramble**, as well as **Prickly Pear** and **Agave** - all plants which thrive in disturbed areas. The woolly-looking patches on some of the **Prickly Pears** are **Cochineal Bugs**, which have been cultivated in the past for the production of the red cochineal dye. The **Agave** is a very spectacular plant with a rosette of enormous fleshy tapering leaves ending in a vicious thorn, and flower stems which can reach to a height of about 10m (33ft). The numerous flat clusters of bright yellow blossoms on each stem attract bird species such as **Canaries, Blue Tits, Spanish Sparrows** and **Linnets** - a single flowering stem providing a pyramid of bird tables one above another. If you are lucky you might see two of the rarest butterflies in the island, the **African Grass Blue** and **Long-tailed Blue**.

SOME NOTABLE PLANTS. *Adiantum capillus-veneris, Aeonium lindleyi, Agave* sp., *Ageratina adenophora, Allagopappus dichotomus, Artemisia thuscula, Arundo donax, Asparagus* sp., *Datura stramonium, Dittrichia viscosa, Dracaena draco* (**Dragon Tree**), *Euphorbia balsamifera (**Tabaiba Dulce**), E.canariensis (**Cardón**), E.obtusifolia (**Tabaiba**), Forsskaolea angustifolia, Kleinia neriifolia, Lavandula* sp., *Lavatera acerifolia, Nicotiana glauca* (**Tree Tobacco**), *Opuntia* sp. (**Prickly Pear**), *Pancratium canariense, Periploca laevigata, Phoenix canariensis* (**Canary Palm**), *Plocama pendula, Ricinus communis* (**Castor Oil Plant**), *Rubia fruticosa, Rubus* sp. (**Bramble**), *Salix canariensis* (**Canary Willow**), *Salvia broussonetii, Schizogyne sericea, Scilla haemorrhoidalis, Sonchus* sp., *Tamus edulis*.

Birds you might see. Buzzard, Kestrel, Barbary Partridge, Herring Gull, Rock Dove, Long-eared Owl, Plain Swift, Hoopoe, Berthelot's Pipit, Grey Wagtail, Robin, Blackbird, Spectacled Warbler, Sardinian Warbler, Blackcap, Chiffchaff, Blue Tit, Raven, Spanish Sparrow, Canary, Goldfinch, Linnet.

Butterflies you might see. African Grass Blue, African Migrant, Bath White, Canary Blue, Canary Speckled Wood, Clouded Yellow, Long-tailed Blue, Lulworth Skipper, Monarch, Painted Lady, Plain Tiger, Red Admiral, Small Copper, Small White.

Dragon Tree

6. *Ladera de Güímar* (Güímar Cliff)

Habitat description

North facing inland cliff in the lower zone, at an altitude of about 500m (1650ft).

Special points of interest

A rich community of endemic plants, with access along a disused water channel that runs across the face of the cliff.

HOW TO GET THERE. *GÜÍMAR* is halfway down the east side of the island, on the old road which is inland from the motorway. See map on page 62. **Bus**: See Appendix. **Car**: Take the C822 south from *GÜÍMAR*. Soon after the road leaves the town it starts to climb across the cliff face. About l/2km past the large white building perched on the cliff edge the road reaches its highest point. At the point where there is a metal electricity pylon on the left, there is an entrance to ar *Mirador don Martín* an old quarry on the right where you can park. This is about 4km (2.5m) from *GÜÍMAR*. Walk up the rough road behind the quarry (suitable for 4-wheel drive vehicles only). About 1/2km up this track it is crossed by electricity wires with wooden poles. Turn right here by a large concrete building and cross the white pumice rubble and old terracing and several old water channels. Turn left at the fourth wooden pole and join the water channel that goes along the face of the cliff. There is a narrow footway alongside the water canal.

PRACTICAL DETAILS. Definitely not a place for small children or anyone who does not like heights; the water channel goes along the cliff, often with a long vertical drop. The accompanying narrow footway is safe (at the time of writing) for at least .5km. It takes 15min to walk from the road to the cliff face. Even as little as a 5 minute exploration along the water channel is worthwhile.

The excursion

The Ladera de Güímar is a dramatic inland north facing cliff; in many places it is more or less vertical. The disused water channel runs along the face of the cliff; in some places it is built into the cliff whilst in others it is supported by concrete piers that have their foothold many metres below. This is a somewhat alarming excursion but is a paradise for the intrepid botanist. Once you find the start of the channel you can't lose your way - it is either forward or back. Three of the large euphorbias grow here - **Cardón**, **Tabaiba** and the less common **Tabaiba Majorera**. There are a few small individuals of **Phoenician Juniper**, a tree that was probably once much more common at this level in the southern lower zone. You may also see small specimens of **Visnea**, a rare tree that is more typical of the laurel forest. A number of other endemic shrubs and herbs grow on the cliffs, many of them with very restricted distribution. These include two species of **Sonchus**; a giant crucifer **Crambe**, and a giant scabious relative

Pterocephalus *(P.dumetorum)*. A rare perennial herb that can be found here is the white-flowered umbellifer **Tinguarra**, a Canary endemic genus. In spring one of the most attractive flowers is a brilliant purple **Pericallis** *(P.lanata)*, but in summer this and many of the other smaller plants are withered and hard to recognise. One that can still be found in summer, in the crevices, is the succulent **Monanthes** *(M.adenoscepes)* with a tiny rosette of crowded leaves; it is known only from this locality, but is a member of a highly successful genus endemic to Macaronesia, with 14 species on Tenerife, several of which hybridize. The **Long-tailed Blue**, which is one of the rarer butterflies on the island, has been recorded from here.

SOME NOTABLE PLANTS. *Aeonium* sp., *Asparagus* sp., *Asphodelus* sp., *Campylanthus salsoloides, Crambe arborea, Euphorbia atropurpurea (**Tabaiba Majorera**), E.canariensis (**Cardón**), E.obtusifolia (**Tabaiba**), Hypericum canariense, Juniperus phoenicea* (**Phoenician Juniper**), *Kleinia neriifolia, Lavatera acerifolia, Monanthes adenoscepes, Opuntia* sp. (**Prickly Pear**), *Pericallis lanata, Periploca laevigata, Pterocephalus dumetorum, Rhamnus crenulata, Rumex lunaria, Salvia canariensis, Sonchus* spp., *Tinguarra cervariaefolia, Visnea mocanera*.

BIRDS YOU MIGHT SEE. Sparrowhawk, Buzzard, Kestrel, Barbary Partridge, Rock Dove, Turtle Dove, Plain Swift, Hoopoe, Berthelot's Pipit, Robin, Blackbird, Spectacled Warbler, Sardinian Warbler, Blackcap, Chiffchaff, Blue Tit, Linnet.

BUTTERFLIES YOU MIGHT SEE. African Migrant, Bath White, Canary Blue, Canary Speckled Wood, Clouded Yellow, Indian Red Admiral, Long-tailed Blue, Monarch, Painted Lady, Red Admiral, Small Copper, Small White.

Hoopoe

7. *Degollada de Cherfe* (Cherfe Pass)

Habitat description

Mountain pass at 1000m (3300ft), with rich dry country scrub vegetation. Spectacular eroded ridges and *barrancos* in a geologically ancient and unspoilt part of the island.

Special points of interest

Rich communities of endemic plants.

HOW TO GET THERE. *DEGOLLADA DE CHERFE* is between *SANTIAGO DEL TEIDE* and *MASCA* in the *TENO* peninsula in the extreme west of the island. **Bus:** See Appendix. **Car:** Take the road west from *SANTIAGO DEL TEIDE*, which goes over the mountains of *TENO* and eventually down to *BUENAVISTA* in the north. Drive 1.5km (1mi) up this steep windy road and park on the ridge.

PRACTICAL DETAILS. No special footwear for the wander we suggest; even a few minutes near your car on the ridge would be rewarding. If you walk further, go well shod and well prepared in general. If you plan to explore the *MASCA* gorge, for instance, you will need a full day: it is a very steep walk down to the sea, and you have to come back up again. (See LANDSCAPES OF TENERIFE.)

The excursion

The Degollada de Cherfe is the highest point of the road between Santiago del Teide and Masca, in the Teno peninsula. This is much higher than the other lower zone localities in this book, and is in fact as high as the laurel forest in Anaga at the other end of the island.

To the right of the parking place it is easy to walk out along the ridge for a little way and explore a nearby gully. In contrast to the other lower zone places that we describe the dominant euphorbia here is **Tabaiba Majorera** with *Tabaiba* only on the lower slopes. Another endemic shrub to be found here is the bushy yellow-flowered umbellifer **Bupleurum**. Of particular interest is a small shrub with narrow sticky leaves which is one of the two species of **Phyllis** on Tenerife: this genus is found on the Canaries and Madeira, but its closest relatives are far on the other side of the equator in southernmost Africa. Also in this area are two white-flowered leguminous shrubs, **Retama** *(R.raetam)* with small flowers and **Chamaecytisus** with larger ones - the latter a tall shrub of the higher parts of the lower zone and the forests. Small plants include a very local species of the labiate genus **Sideritis**, with densely white-felted stems; you may also see its close relative **Salvia** with broader leaves and branched flower heads with pale mauve flowers. There is also a yellow thistle **Carlina**. Succulents include a tall, unbranched **Aeonium**, a species of **Greenovia** with rosettes of leaves at ground level, and a tiny **Monanthes** in cracks.

If you turn over rocks here or rather lower down on either side of the ridge you may come across the large cricket, *Gryllus bimaculatus*, with a dark body and two pale spots at the base of the wings; it is widespread in Tenerife but also in many other parts of the Old World. However don't keep your eyes down all the time since this is one of the places where you might find the **Rock Sparrow**.

SOME NOTABLE PLANTS. *Aeonium* sp., *Bupleurum salicifolia, Carlina salicifolia, Chamaecytisus proliferus, Echium aculeatum, Euphorbia atropurpurea (***Tabaiba Majorera***), E.obtusifolia (***Tabaiba***), Foeniculum vulgare, Greenovia* sp., *Kleinia neriifolia, Monanthes* sp., *Phyllis viscosa, Retama raetam, Salvia canariensis, Sideritis* sp.

BIRDS YOU MIGHT SEE. Buzzard, Kestrel, Barbary Partridge, Rock Dove, Plain Swift, Berthelot's Pipit, Spectacled Warbler, Sardinian Warbler, Chiffchaff, Blue Tit, Raven, Rock Sparrow, Canary, Linnet.

BUTTERFLIES YOU MIGHT SEE. Bath White, Canary Blue, Canary Speckled Wood (?), Clouded Yellow, Painted Lady, Red Admiral, Small Copper, Small White.

Kleinia neriifolia

8. *Barranco del Infierno* (Hell's Gorge)

Habitat description

A deep ravine or gorge high up in the lower zone - about 400m (1300ft) - terminating in vertical cliffs and a waterfall.

HOW TO GET THERE. *BARRANCO DEL INFIERNO* is in the southwest of the island near *ADEJE*, which is 11km (7mi) north of *LOS CRISTIANOS*. **Bus**: See Appendix. **Car**: The town of *ADEJE* is built on cliffs overlooking the gorge, and you can walk into it from this town. Go up the main tree-lined street and turn left at the top. Follow the road round a sharp right bend and up hill (barely 1/2 km) until the road stops. There is a small car park on the left about three houses from the end. Walk down the signposted path immediately beyond the last house on the right.

PRACTICAL DETAILS. If you can, go early in the day; it can be very hot indeed later on. Normal walking shoes should be sufficient, as the track is quite easy. Allow three hours for a leisurely walk to the end of the track and back.

The excursion

Barranco del Infierno comes into sight immediately you leave the road beyond the parking place and from there you also have a dramatic view down the *barranco* towards the sea. The first stretch of the track runs along the side of the *barranco* and slowly converges with its floor. There are large numbers of beehives near the start of the walk. Some of these are the old-fashioned type made out of palm tree trunks, and some are the more modern box type. If you come here in the spring or early summer it will be immediately obvious why this is such a good place for bees; it is a dazzling mass of flowering shrubs. One of the more interesting plant species is the endemic **Justicia**; it is a small shrub with whitish hooded flowers. You may also see the composite **Allagopappus**, a shrub with narrow toothed sticky leaves and dense, flattish heads of yellow flowers: this genus is endemic to the Canaries. The endemic euphorbias *Cardón* and *Tabaiba* are dominant here and although most of the other plants are also native to Tenerife, there are a few introduced ones; these include the **Prickly Pear**, **Tree Tobacco** and the occasional **Mulberry** and **Fig**. Some of the **Prickly Pear** plants have **Cochineal Bugs** *(Dactylopius coccus)* on them, looking like white dusty patches: this is a species of scale insect introduced to the islands from Mexico. If you squeeze one you will find the deep red cochineal dye on your fingers: this used to be one of the main exports of the Canary Islands before the days of aniline dyes, and it is still produced commercially on Lanzarote, but only in a very small way on Tenerife.

Some of the time you will be walking close to water canals which take the water from higher up the gorge down to the towns and agricultural areas below: that is why the gorge is almost dry lower down. The path is fairly clearly marked with yellow arrows, but there is one place where people frequently go wrong. When you first see a track below you going fairly steeply down into the *barranco*, just after you have rounded a sharp corner, watch out for the right turn onto this track - about 40m (44yds) ahead. You have to cross a water canal (there is an inconspicuous yellow arrow on its side) and not continue up to your left.

As you descend towards the damper bottom of the gorge you will find that the vegetation changes; there are thickets with **Bramble**, **Bracken**, **Maidenhair Fern** and also the **Canary Willow.** This willow, which can be found in damp places in several parts of Tenerife, sometimes provides an extraordinary entomological spectacle, since in early spring the trees are often infested by thousands of caterpillars of the endemic moth *Yponomeuta gigas*; the caterpillars weave a communal web that may entirely envelop the tree, thus giving them protection while they consume the leaves. In this area a few non-native trees have been planted - Almond, Walnut and Pear.

You will come to an area with a network of pipes and canals of different ages, and also a small dam; this *barranco* is one of the few places on the island where there is still a little natural running water all the year. Dragonflies are very common here, and in summer you are certain to see some of the ten species which occur on the island. *Trithemis arteriosa* is a particularly abundant dragonfly; it is an African species with a bright red male and black and yellow female. In the pools you may find tadpoles and several kinds of water beetles; there are also waterboatmen (Corixidae) and also the endemic backswimmer *Notonecta canariensis* and pondskater *Velia lindbergi*. Another insect you might see is a large (about 6cm long) spotted African mantid, *Blepharopsis mendica*.

After this the *barranco* becomes quite narrow and the path stops where the cliff walls finally converge. Here there is a waterfall, which even in the middle of summer has water trickling down the rock face. If you look upward to the high cliffs along this stretch you may be able to see wild **Dragon Trees** clinging to the rocks. These are not the huge trees growing in sheltered conditions in the town parks, but scrawny isolated specimens growing under harsh natural conditions. Even higher up you can see pine trees at the lower limit of the forest zone. High up on some of the slopes, where they run up to the bottom of vertical cliffs, you may be able to see holes belonging to **Cory's Shearwater**. These birds come inland at dusk to their nests, sometimes in old rabbit holes, in *barrancos* throughout the island; their weird cries at night give rise to many a strange legend, and may even account for the name of this *barranco*. The whole valley is a very good place for birds; there are **Sardinian Warblers** and **Blackcaps** singing tantalizingly hidden in the dense thickets of shrubs, **Barbary Partridges** calling from the slopes, **Kestrels** hunting for the **Canary Lizard** and the occasional **Buzzard** calling overhead.

SOME NOTABLE PLANTS. *Adiantum capillus-veneris* (**Maidenhair Fern**), *Aeonium urbicum, Allagopappus dichotomus, Campylanthus salsoloides, Ceropegia fusca, Cistus monspeliensis, Dittrichia viscosa, Dracaena draco* (**Dragon Tree**), *Echium aculeatum, Euphorbia atropurpurea (**Tabaiba Majorera**), E.canariensis (**Cardón**), E.obtusifolia* (**Tabaiba**), *Ficus carica* (**Fig**), *Forsskaolea angustifolia, Justicia hyssopifolia, Kicksia scoparia, Kleinia neriifolia, Launaea arborescens, Lavandula sp., Marcetella moquiniana, Morus nigra* (**Mulberry**), *Nicotiana glauca* (**Tree Tobacco**), *Opuntia sp.* (**Prickly Pear**), *Pancratium canariense, Periploca laevigata, Plocama pendula, Pteridium aquilinum* (**Bracken**), *Rubus sp.* (**Bramble**), *Rumex lunaria, Salix canariensis* (**Canary Willow**), *Sideroxylon marmulano.*

BIRDS YOU MIGHT SEE. **Cory's Shearwater, Sparrowhawk, Buzzard, Kestrel, Barbary Partridge, Rock Dove, Turtle Dove, Long-eared Owl, Plain Swift, Berthelot's Pipit, Grey Wagtail, Robin, Blackbird, Blue Tit, Raven, Canary, Linnet, Sardinian Warbler, Blackcap, Chiffchaff.**

BUTTERFLIES YOU MIGHT SEE. **African Migrant, Bath White, Canary Blue, Canary Speckled Wood, Clouded Yellow, Indian Red Admiral** (rare), **Monarch, Painted Lady, Plain Tiger, Red Admiral, Small Copper, Small White.**

Plocama pendula

LAUREL FOREST - *MONTEVERDE*
(Excursions 9-12)

9. *Anaga* beyond *El Bailadero*

Habitat description
The most extensive surviving area of laurel forest on Tenerife, at about 800m (2600ft), with examples of both *laurisilva* and *fayal-breza*l (see ECOLOGICAL ZONES, Section 1).

Special points of interest
A unique opportunity to walk through a relic of the type of forest that covered much of southern Europe around 20 million years ago. A part of this area is now administered by the regional government and subject to strict control over use.

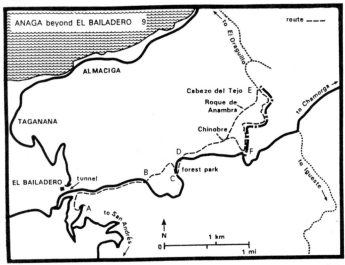

HOW TO GET THERE. *EL BAILADERO* is on the mountain ridge in the centre of the *ANAGA* peninsula in the extreme east of the island; it is not a village but just a group of restaurants. **Bus**: See Appendix. **Car**: Drive either from *LA LAGUNA* or from *SAN ANDRES* to *EL BAILADERO* and from there to any of the points B to E along the road that winds along the central ridge of the *ANAGA* peninsula. If you drive up the very steep road from *SAN ANDRES* take care that you do not go through the tunnel in the mountains, but instead take the turning left a short distance before it, and then turn right again after 2km (1.2mi).

Pᴙᴀᴄᴛɪᴄᴀʟ ᴅᴇᴛᴀɪʟs. This is a region of very high humidity. Whatever the weather down below, up here you are likely to be in cloud, and it may well be raining. It also can be quite cold. So, even in July or August, take rainwear and a spare warm layer. Because of the humidity, the footpaths can be very slippery, although good steps have been cut for the steeper parts: sturdy footwear is needed. We make no excuse for giving detailed directions for these walks; it can be a very confusing area to find one's way in, and available maps are totally inadequate. If you have only a short time we would recommend walking from Points B to C (see Map) and back along the road; this would take about 1 1/2hrs. From Points D to F and back to your car at D would take about 1 1/2hrs. From D to E and then along the forest track to F and back along the road to D could take over 2 hrs. If you go by bus and walk from Points A to E it would take about 2 1/2hrs. You will need to decide how to return. Walking down to *EL DRAGUILLO* in the north and then on to *ALMACIGA* where there is a bus would take an additional 2hrs. An alternative way would be to walk east along the road from Point F for about 1.5km (1mi) and find the track that leads down to *IGUESTE DE SAN ANDRES* in the south (or follow Walk 31 in LANDSCAPES OF TENERIFE). Although the logistics may seem a bit daunting, we find walks in *ANAGA* enormously rewarding and well worth the effort needed to work out a practicable route.

The excursion

On this part of the Anaga peninsula there remains one of the largest surviving and relatively undisturbed areas of subtropical evergreen forest. There are several well marked forest trails and we describe just three sections which can either be followed as separate short walks or as part of a much longer walk.

Points A - B. When you leave the bus (at A), cross the road and climb straight up the hill through evergreen forest on a track which starts just before the tunnel entrance; it is marked on a yellow sign *"EL BAILAD-ERO"*. In about 5 minutes you will come to a T junction in the path and you go right (signed to *"EL BAILADERO"*). Many of the plants that occur later in this walk are also found here; it is a particularly good place for the **Canary Bellflower**. After about another five minutes of steep climbing you come to the high road from La Laguna, where you turn right. You have now done the only significant climb of this walk! Here at El Bailadero there is a *mirador* with a magnificent view out to the valley of Taganana in the north, and a couple of restaurants. The bright lights around these isolated buildings tend to attract insects at night, and some of these stay on the adjacent walls during the following day; you may find a variety of species there, including the very large **Convolvulus Hawkmoth**, with pink bands around its abdomen. From El Bailadero to B on the map will take around 30 minutes. The road runs for much of the way along the very exposed high mountain ridge. The clouds (or rain) whip across almost horizontally from the north. Even when it hasn't been raining the road is often wet below the trees. This is due to the continual dripping of water that has condensed out of the clouds onto the leaves. This condensation in the forest is a very important source of water for the island.

The most conspicuous tree along this ridge is the *Tejo*. This tree heath has bark which tends to peel off in strips, unlike the other tree heath, *Brezo*, which has finely grooved bark, and does not grow in such exposed situations. Under the bark of *Tejo* is a good place to search for beetles; a black endemic weevil *Laparocerus undulatus* lives here, and also the black tenebrionid *Nesotes conformis*.

The road passes through several cuttings and the rock faces are colonized by such plants as the endemic **Sonchus** species and especially by the succulent **Aeonium**, which seems to thrive in the damp conditions near the ridge, and whose dense rosettes give a dramatic effect when viewed through swirling mist. Large specimens have leafy flowering stems up to 60cm (2ft) high, ending in spectacular cones of yellow flowers. The litter of leaves around **Aeonium** plants are good places to find some of the special insects of the laurel forest. One of these is a delicate cockroach *Phyllodromica bivittata*, which has winged males but wingless females; another is an endemic earwig *Guanchia cabrerae*. Under loose rocks there is a chance of finding the long-legged, high-speed centipede *Scutigera coleoptrata*, although this is more often seen around houses; it is a species that occurs in southern Europe and is established in the Channel Islands but not in mainland Britain; its last (fifteenth) pair of legs are especially long and look very like the antennae at the head end, giving the animal a striking push-me-pull-you appearance.

Points B - C. When you reach point B you will see some steep earth steps cut in the bank on your left, signposted to *"EL PIJARAL"*. Before you go up these steps it is worth looking at the rock face which was exposed when this road was cut, where you can see two conspicuous diagonal dykes and also a smaller one just at the beginning of the steps: dykes are described in Section 1 - GEOLOGY. Along either side of each of the dykes is a reddish layer where the surrounding soil was baked by the molten rock. Low down near road level you can see successive layers of pyroclastic deposits - airborne material from volcanic explosions. Although you could walk from

B to C in 30 minutes, you are likely to take longer, especially if you stop to hunt for plants and take photographs. It is not a walk to hurry over. Nearly all the plants that surround you are restricted to the Canary Islands, or at most to the Canaries and the other eastern Atlantic archipelagos.

When you climb the steps at Point B you come first to an area of *fayal-brezal* (exposed ridge type, see ECOLOGICAL ZONES, Section 1), with **Myrica** and the tree heath *Tejo* dominating, but also with some **Laurus, Small-leaved Holly, Rhamnus** and **Visnea**. You soon enter the *laurisilva* with its richer community of trees. The track forks at about 10-15 minutes from the road; turn left here - there is no signpost. About 5 minutes after this at another unsignposted fork, take the right. Stay with the main track which is generally slightly banked at the sides. You can then walk on through the forest until you reach the road at Point C. The track takes you through subtropical rain forest, with trees and ferns continually dripping and mats of moss and lichens hanging on the trees, frequently with ferns growing out of them. There are at least a dozen species of more or less laurel-like trees in this forest; you can find **Laurus, Prunus, Small-leaved Holly, Large-leaved Holly, Heberdenia, Persea, Rhamnus, Visnea**, and **Picconia** (all briefly described in Section 3). In addition there is the tree heath *Brezo* and a common and fairly easily recognized **Viburnum**. The latter is really a large shrub which tends to form an understory in the forest; it has the dense white flower clusters characteristic of the viburnums and their relatives the elders. You will also find many ferns, of which the enormous **Woodwardia** is the most impressive but an endemic buckler-fern, **Dryopteris**, is more widely distributed.

A number of other characteristic plants of the *laurisilva* may be found near here. A particularly interesting one is a labiate of the genus **Bystropo-gon**: it is a tall aromatic shrub with oval leaves and clusters of small, round, pink or white flower heads. Members of this genus occur only in the Canaries, Madeira and western South America! One of the endemic composites is a **Pericallis** (*P.cruenta*) growing up to 1m high, with broad, shallow-lobed leaves, glossy green above and white and woolly below; the clusters of flowers are white with yellow centres. In the summer the leguminous shrub **Teline** is in full bloom with showy yellow flowers. You will need to come in spring to see the orange flowers of the **Canary Bellflower**, which is a giant campanula whose closest relatives are found on mountains in Africa; it is a scrambling vine up to 3m (10ft) long, and has opposite, long, triangular leaves with strongly toothed margins; it dies back to a tuberous root each year. If you shake the flowers of this vine, drops of sweet nectar will fall.

Another attractive scrambler in this area is a **Convolvulus** (*C.canariensis*), with densely hairy leaves and pale blue or mauve flowers. Among the herbs one that is noticeable is **Ixanthus**, an endemic gentian relative which grows up to 75cm (2.5ft) tall; it has almost stalkless opposite leaves with a few parallel veins, and a branched flowering head with attractive yellow flowers. You may also find a cranesbill, **Geranium**, locally called *Pata de Gallo* (Cock's foot). Not much light penetrates the forest canopy, but where it does, you may see even in midsummer the brilliant orange flowers of **Isoplexis**: this genus is found only in Madeira and the Canaries, and is a fairly close relative of the foxgloves. You will find a little running water - a rarity in Tenerife - in one or two places along this walk, with mosses and liverworts growing on the damp rocks. When you come back on to the road at Point C there is a yellow sign pointing back up the way you have come saying *"EL BAILADERO"* and one pointing along the road to your left saying *"CABEZO DEL TEJO"*. (If for any reason you want to do this walk 'backwards' and to leave a car near here, there is a pull-off for about 3 cars 50m (55yds) back along the road to your right at this point.)

Points D - E. From Point C walk along the road to the left for about 9 minutes until you come to Point D which is an area with picnic tables, signposted *"PARQUE FORESTAL"*. Don't go along the broad track to your left but take the smaller left track up past the picnic tables with a yellow sign saying *"CABEZO DEL TEJO"* and *"CHINOBRE"*. You will need 1hr to walk from D to E, but it is worth giving it more time if you can. After about 20 minutes you come to a left turn signed to *"CHINOBRE"*. One minute up this path will bring you to this magnificent look-out place (if there is no cloud). Notice the very corky bark of **Myrica** growing here - a good diagnostic feature of this tree when reasonably old. Return to the main path and in about one minute it will fork. Take the left fork marked *"CABEZO DEL TEJO"*. (If you take the right fork at this point you will come down to the road at Point F, which you might want to do if you have left your car at Point D and only have time for a short walk). Over halfway along this walk you will come to the spectacular landmark, Roque de Anambra. If you have been walking in cloud, the high vertical face of this rock may suddenly loom up only a few yards in front of you.

The tree community in this section of the forest is similar to that described for Points B - C. As you walk along you might see part of the ground littered with freshly fallen leaves. On closer examination you will see that the twigs have been bitten through. This is the work of the introduced **Roof Rat**. It attacks mainly **Heberdenia** and **Laurus**, but also **Persea** and **Picconia**. Only small parts of the twigs seem to be eaten, but it is possible that they are the most nutritious ones. It is not certain whether or not this is having a detrimental long-term effect on these species of trees.

If you search the leaves of saplings in the forest you may find the omnivorous slug/snail *Insulivitrina lamarcki*; although this has no external shell it is really a type of snail and is found only on the Macaronesian islands. An interesting insect to be found here is the large endemic cranefly *Tipula macquarti*, which has tiny non-functional wings. The leaf litter in the laurel forest is very rich in invertebrates - although in many areas it is often raked up and carried away. Among the somewhat surprising animals to be found in it is the amphipod crustacean *Orchestia chevreuxi*, a terrestrial member of a group more typical of the sea shore.

It is generally quiet in the forest, but if you hear a noisy bird flying up from the trees above you, it is likely to be a **Bolle's Laurel Pigeon**, a species which occurs only on Tenerife, La Palma, Hierro and La Gomera. You are unlikely to get more than a glimpse from below because it typically feeds high up on the fruit of various trees of the laurel forest. This is not a good place to look for the **White-tailed Laurel Pigeon**, which is rare on Tenerife, although you might have a chance of seeing it in some other places (see *MONTE DEL AGUA* and *CHANAJIGA*, EXCURSIONS No.11 and 12). You will also see along this section of the walk some especially fine specimens of the tree heath *Tejo*, which here is growing up to 6m (20ft); it is more typically only a tall shrub of the high ridges. You will notice that the really tall trees of the *laurisilva* are all in the more sheltered and less steep valleys. On the ridges the trees are much shorter, forming a *fayal-breza*l community (exposed ridge type, see ECOLOGICAL ZONES, Section 1), usually with **Tejo** and **Myrica** dominating, and with some **Heberdenia, Small-leaved Holly, Large-leaved Holly** and **Prunus**.

Points E - F. This is an easy stretch of the walk, along a forest road, and it will take about 40 minutes. In areas where the undergrowth has been cleared along this road, you can get views out over the forest and it could be worth stopping for a while, in the hopes of getting a glimpse of the **Bolle's Laurel Pigeon** as it flies above the trees below you.

SOME NOTABLE PLANTS.

ALONG THE ROAD: *Aeonium canariense, Ageratina adenophora, Bystropogon canariensis, Hypericum* sp., *Lonicera etrusca., Opuntia* sp. (**Prickly Pear**), *Pericallis cruenta, Phyllis nobla, Pteridium aquilinum* (**Bracken**), *Rubus* sp. (**Bramble**), *Rumex lunaria, Sonchus* sp., *Tamus edulis, Teline canariensis*.

TREES IN THE FOREST: *Apollonias barbujana, Erica arborea (**Brezo**), Erica scoparia* (**Tejo**), *Gesnouinia arborea* (rare), *Heberdenia excelsa, Ilex canariensis* (**Small-leaved Holly**), *Ilex perado* (**Large-leaved Holly**), *Laurus azorica, Myrica faya, Ocotea foetens, Persea indica, Picconia excelsa, Pleiomeris canariensis* (rare), *Prunus lusitanica, Rhamnus glandulosa, Sambucus palmensis* (rare), *Viburnum tinus, Visnea mocanera* (rare).

SOME OTHER PLANTS IN THE FOREST: *Adenocarpus foliolosus, Bencomia caudata* (on cliffs, rare), *Bystropogon canariense, Canarina canariensis* (**Canary Bellflower**), *Cedronella canariensis, Convolvulus canariensis, Dracunculus canariensis, Dryopteris oligodonta, Geranium canariense, Hypericum inodorum, Isoplexis canariensis, Ixanthus viscosus, Myosotis latifolia, Pericallis cruenta, Phyllis nobla, Pteridium aquilinum* (**Bracken**), *Ranunculus cortusifolius, Semele androgyna, Teline canariensis, Urtica morifolia, Woodwardia radicans.*

BIRDS YOU MIGHT SEE. **Sparrowhawk, Buzzard, Kestrel, Woodcock, Rock Dove, Bolle's Laurel Pigeon, Plain Swift, Grey Wagtail, Robin, Blackbird, Sardinian Warbler, Blackcap, Chiffchaff, Goldcrest, Blue Tit, Raven, Chaffinch, Canary.**

BUTTERFLIES YOU MIGHT SEE. **American Painted Lady, Bath White, Brown Argus, Canary Blue, Canary Speckled Wood, Cardinal, Cleopatra, Clouded Yellow, Indian Red Admiral, Large White, Meadow Brown, Painted Lady, Red Admiral, Small Copper, Small White.**

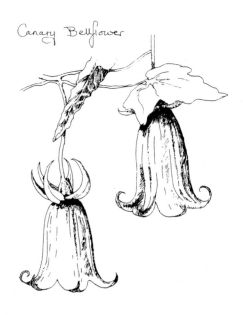

Canary Bellflower

10. *Pico del Inglés* (Englishman's Peak)

Habitat description

A *mirador* overlooking laurel forest and a nearby path down a steep slope through the forest.

Special points of interest

A good spot to see the endemic **Bolle's Laurel Pigeon**.

HOW TO GET THERE. The *mirador* of *PICO DEL INGLES* is towards the western end of the *ANAGA* peninsula. **Bus:** See Appendix. **Car:** In *LA LAGUNA* find the road to *"LAS MERCEDES"* and *"PUNTA DEL HIDALGO"*. About 3km (2mi) along this road you come to a village called *"LAS CANTERAS"*, where the road forks. Take the right fork marked *"TF114 MONTE DE LAS MERCEDES"*. From here follow signs to *"PICO DEL INGLES"*, and *"TF1123"*. Pass the *mirador CRUZ DE CARMEN* and ignore left turns marked *"CF4"* and *"TF12"*. *PICO DEL INGLES* is nearly 8km (5mi) from *LAS CANTERAS*.

PRACTICAL DETAILS. The path is very slippery.

The excursion

If you wish to visit the laurel forest we recommend further east in Anaga (see *ANAGA* BEYOND *EL BAILADERO*, EXCURSION No.9), but if you do not have time for that a visit to Pico del Inglés is well worth while. Immediately around the parking area there are several of the endemic trees and a short walk to the *mirador* will enable you to look out over a large expanse of forest. Across the trees to the west you can see bare hillsides which were, in the past, also covered with laurel forest. The steep ridges and gullies of this area show much more clearly there; nearby, the contours are softened by the forest where the trees of the *laurisilva* in the gullies grow taller than those of the ridge-top *fayal-breza*l (see ECOLOGICAL ZONES, Section 1).

One reason for choosing this area is that it is probably the best spot in the east of the island for the visitor to see the endemic **Bolle's Laurel Pigeon**. In the morning, just as the sun reaches the tree tops, and before the crowds of visitors arrive at this popular view point, is the best time to watch for pigeons as they fly across the tops of the trees far below you. They feed on the berries of **Laurus** and other species of native trees. The other endemic pigeon - the **White-tailed Laurel Pigeon** - is rare on Tenerife and extremely unlikely to be seen at this end of the island. Blackbirds are strikingly common in this area, as well as in the dryer valleys lower down.

If this is your only visit to the laurel forest region it is worth walking down into part of the forest here, on a path starting at a point about 0.5km back along the road. On the opposite side of the road from Bodegón Cruz de Afur is a path signposted *"LA ASOMADA"*. This track is rather steep and slippery, but you will find examples of many of the trees typical of the laurel forest, although the area has been cut for timber in the past. After 10 - 15 minutes walking down, the track levels out and after about 3 minutes more you come to the entrance of a water *galería* with a locked metal door in the rock face; nearby is a water drip where you may see forest birds drinking. There is another very slippery descent and a second *galería* entrance further on.

One of the less obvious but interesting trees grows close to the upper *galería* entrance: it is **Persea**, a member of the laurel family which is now found only in Macaronesia but with relatives in America and East Africa. Distinguishing it from the other laurel-like trees is made fairly easy by the fact that most individuals have a few bright orange-red old leaves. Also growing close to the path is **Rhamnus**, a small tree with oval serrated leaves that have several prominent round glands between the bases of the veins; other species of this genus grow in the lower zone and the high mountain zone.

Scrambling shrubs and vines are noticeable here. Apart from the local ivy, **Hedera**, which is now considered to be the same species as the mainland European form, there are two less familiar vines. One of them, **Tamus**, is related to the yams, with large underground tubers; it has alternate, pointed, heart-shaped leaves, flowers on a spike and reddish-orange berries. The other, **Smilax**, is a member of the lily family; it also has reddish fruits, but is distinguishable from **Tamus** by the arrangement of its flowers, which are in a flat cluster, and by the thorns on its stems.

The two tree heaths are both conspicuous along this path, but in rather different situations. In the lower, less steep areas are fine specimens of *Brezo* with hard, finely-grooved bark. Higher up, however, and on the steeper and more exposed slopes, you will see more of the smaller *Tejo*, characterized by its more geometric leaf arrangement and the tendency for its bark to peel off in strips.

SOME NOTABLE PLANTS. *Adenocarpus foliolosus, Erica arborea (**Brezo**), E.scoparia (**Tejo**), Heberdenia excelsa, Hedera helix, Ilex canariensis (**Small-leaved Holly**), I.perado (**Large-leaved Holly**), Ixanthus viscosus, Laurus azorica, Myrica faya, Persea indica, Picconia excelsa, Pleiomeris canariensis, Prunus lusitanica, Rhamnus glandulosa, Smilax canariensis, Tamus edulis, Teline canariensis, Viburnum tinus.*

BIRDS YOU MIGHT SEE. **Sparrowhawk, Buzzard, Kestrel, Woodcock, Rock Dove, Bolle's Laurel Pigeon, Turtle Dove, Plain Swift, Berthelot's Pipit, Grey Wagtail, Robin, Blackbird, Sardinian Warbler, Blackcap, Chiffchaff, Goldcrest, Blue Tit, Chaffinch, Canary.**

BUTTERFLIES YOU MIGHT SEE. **American Painted Lady** (rare), **Bath White, Brown Argus, Canary Blue, Canary Speckled Wood, Cardinal, Cleopatra, Clouded Yellow, Indian Red Admiral, Large White** (rare), **Lulworth Skipper, Meadow Brown, Painted Lady, Red Admiral, Small Copper, Small White.**

Laurus azorica

11. *Monte del Agua*

Habitat description
Laurel forest, 1000m (3300ft).

Special points of interest
Undisturbed areas of laurel forest and one of the best opportunities for seeing both **Bolle's Laurel Pigeon** and the **White-tailed Laurel Pigeon**.

HOW TO GET THERE. *MONTE DEL AGUA* is in the *TENO* peninsula between *SANTIAGO DEL TEIDE* and *BUENAVISTA*. **Bus**: See Appendix. **Car**: About midway between the towns of *SANTIAGO DEL TEIDE* and *EL TANQUE* is the small village of *ERJOS DEL TANQUE*. Just south of the village, opposite the *casa forestal* (forestry office) is the start of an unpaved road (unsignposted) which leads into the forest region, and eventually down to join the road between *SANTIAGO DEL TEIDE* and *BUENAVISTA*.

PRACTICAL DETAILS. Easy track for walking on, so no special footwear required. It would take about 1hr to walk from *ERJOS DEL TANQUE* to Emmerson's lookout.

The excursion
Monte del Agua is a forested mountainous region at the head of a huge valley - Barranco de los Cochinos - that cuts deep into the north of the Teno massif. Whereas in Anaga there are several forest trails that are easy to follow, here there are very few; the forest road, however, gives easy access to this fascinating area. After leaving the main road at Erjos del Tanque the unpaved road passes through a *fayal-breza*l scrub forest, with **Adenocarpus**, *Brezo*, **Myrica** and **Bramble** for about 1km. Then it enters an area of *laurisilva* with steep slopes densely covered with laurel-like trees. In some of the gullies and dells there are fine large trees, often with massive growths of lichens on their trunks. The commonest species are **Laurus, Myrica, Brezo, Arbutus, Persea**, and the **Small-leaved Holly**, with **Viburnum** forming the main understory. Interesting smaller shrubs include **Isoplexis** and **Echium**, and you will also see **Hypericum, Rumex**, a pink-flowered **Cistus** and **Geranium**.

After about 3.5km (2mi) there is a water tap where a pool of water accumulates on the road; this is an excellent place to observe the forest birds, especially early in the morning. The road runs in and out of side valleys, along the side of the main *barranco*. After 4.5km (3mi) you come to a conspicuous rock standing up on the right of the road on a left-hand bend. There is a green painted rain gauge on the top. This rock is known to many bird watchers as "Emmerson's lookout" after the English ornithologist who lives on the island, who must have shown dozens of visitors where to get a view of **Bolle's Laurel Pigeon**, and with luck also the **White-tailed**

Laurel Pigeon. The rock is on the edge of the Barranco de los Cochinos and overlooks an extensive area of forest where the pigeons can be seen - especially early in the morning - flying over the tops of the trees in the valley below. **Bolle's Laurel Pigeon** is now considered to be a Canary Island endemic species; there is a very closely related species on Madeira and the two forms have sometimes been treated as subspecies of a single species. It is like a Woodpigeon, but lacks the white wing bars and the white patch on the side of the neck; the tail has a pale subterminal band, beyond which is a dark bar. The **White-tailed Laurel Pigeon** is confined to the western Canary Islands and is now rare on Tenerife. In flight it can be distinguished from **Bolle's Laurel Pigeon** by its browner upperparts and by the tail, which becomes paler towards the tip, without any dark terminal bar. Both pigeons feed primarily on the fruits of the trees in the forest.

There are many interesting insects in the laurel forest, including the rare **Queen of Spain Fritillary**. On the ground you might see a large beetle *Carabus faustus*, which is black with brilliant green highlights. On the trees you can find the green predatory bush cricket *Calliphona koenigi*, a large insect (body length over 2.5cm) with reduced wings, which is known only from the northern side of Tenerife. It does not fly but its black and red hind wings can be exposed suddenly to startle potential predators. This species has relatives with full wings on other islands in Macaronesia. Another bush cricket that can be found on the leaves of the *laurisilva* trees is *Canariola nubigena*. It is much smaller, wingless, and with antennae four times its body length; the females are brown but the males have marbled grey bodies with patches of other colours. Among the spiders of this area is the large yellowish clubionid *(Cheiracanthium pelasgicum)*, a member of a genus in which many species are known to be poisonous.

SOME NOTABLE PLANTS. *Adenocarpus foliolosus, Arbutus canariensis, Cistus symphytifolius, Echium* sp., *Erica arborea* (**Brezo**), *Geranium canariensis, Globularia salicina, Heberdenia excelsa, Hypericum inodorum, Ilex canariensis* (**Small-leaved Holly**), *Isoplexis canariensis, Laurus azorica, Myrica faya, Ocotea foetens, Persea indica, Picconia excelsa, Ranunculus cortusifolius, Rubus* sp. (**Bramble**), *Rumex lunaria, Viburnum tinus, Visnea mocanera.*

BIRDS YOU MIGHT SEE. **Sparrowhawk, Buzzard, Kestrel, Barbary Partridge, Woodcock, Rock Dove, Bolle's Laurel Pigeon, White-tailed Laurel Pigeon, Turtle Dove, Long-eared Owl, Plain Swift, Berthelot's Pipit, Robin, Blackbird, Sardinian Warbler, Blackcap, Chiffchaff, Goldcrest, Blue Tit, Raven, Chaffinch, Canary, Linnet.**

BUTTERFLIES YOU MIGHT SEE. **American Painted Lady, Bath White, Canary Blue, Canary Speckled Wood, Cardinal, Cleopatra, Clouded Yellow, Indian Red Admiral, Lulworth Skipper, Meadow Brown, Painted Lady, Queen of Spain Fritillary, Red Admiral, Small Copper, Small White.**

12. *Chanajiga*

Habitat description
Laurel forest on very steep east facing cliffs above Los Realejos, at 1200m (3900ft).

Special points of interest
We include this site because it is within easy distance of Puerto de la Cruz, and is a good place to see **Bolle's Laurel Pigeon** and the rare **White-tailed Laurel Pigeon**; it is also rich in butterflies.

HOW TO GET THERE. *CHANAJIGA* picnic area is in the north of the island, on the western rim of the *OROTAVA* valley. **Bus**: See Appendix. **Car**: The *Cabildo* map is confusing (and wrong in places) so we give a lot of details here to help you find your way, although it is well signposted. From *OROTAVA* go uphill and follow the sign saying *"PARQUE NACIONAL DEL TEIDE"*. The road is steep and winding. There is a sudden but well marked junction with a sign to the right to *"BENIJOS"* and *"LOS REALEJOS"*. After 4.6km (3mi) there is a left turn marked obscurely on the side of a house on the left *"PISTA FORESTAL, LLANADAS, LAS CANADAS"*. Take this paved road and after 1km pass through the unsignposted village of *LLANADAS*. Keep right at a sign saying *"ZONA RECREATIVA CHANAJIGA"*. You will see the steep wooded cliffs ahead of you, with two forest roads making unsightly marks along them: you are heading for the lower one. The road continues to be well signposted and at about 5.5km (3.4mi) from *LLANADAS* you come to a T-junction where the tarmac ends; turn right and you will find plenty of parking space.

PRACTICAL DETAILS. An easy walk along a level forest road, which continues down to *"ICOD EL ALTO"* (See WALK No.14 in LANDSCAPES OF TENERIFE).

The excursion
Chanajiga forest picnic area is situated on the steep slopes at the west of the huge Orotava valley in the north of the island; it can be hard to find if you depend upon maps. This area is frequently under cloud, so try to choose a clear day for a visit. On these slopes are remnants of the laurel forest that once extended along the whole of the north side of the island, and a short walk from the picnic site will bring you to one of these areas. You will be walking along the forest road that you saw earlier from below. At first you pass through an area where pines have been planted. These are mainly the introduced **Monterey Pine**, with a few **Canary Pine** further on. There is *fayal-brezal* undergrowth dominated by **Myrica**, along with **Brezo**, a reminder of the ancient laurel forest; **Cistus** and **Viburnum** are also conspicuous along here. After about 15 minutes of walking you come to an area of laurel forest. This is perhaps not the best area to visit if your primary interest is in the ancient forest; this is because the slope is incredibly steep

and the only practical access is along the forest road. However, several trees typical of the laurel forest can be seen here (see ECOLOGICAL ZONES, Section 1 and *ANAGA* BEYOND *EL BAILADERO*, EXCURSION No.9) with **Small-leaved Holly, Laurel** and **Myrica** dominating, while in more open patches you will see a very large white-flowered **Echium**. Perpetually damp areas on the rock faces add to the variety of habitat, and together with the flowers along the track edge, attract numerous butterflies, especially **Canary Speckled Wood**.

The forest road is a useful vantage point for looking out over the forest below you. If you are patient you will almost certainly see one or both of **Bolle's Laurel Pigeon** and **White-tailed Laurel Pigeon** making short flights across the tree tops - except of course when the whole area is blanketed in cloud.

SOME NOTABLE PLANTS. *Adenocarpus foliolosus, Aeonium* sp., *Bencomia caudata, Carlina salicifolia, Cedronella canariensis, Chamaecytisus proliferus, Cistus monspeliensis, Cistus symphytifolium, Echium giganteum, Erica arborea (**Brezo**), Geranium canariense, Hypericum inodorum, Hypericum reflexum, Ilex canariensis* (**Small-leaved Holly**), *Laurus azorica, Myosotis latifolia, Myrica faya, Pericallis cruenta, Persea indica, Phyllis nobla, Pteridium aquilinum* (**Bracken**), *Rubus* sp. (**Bramble**), *Scrophularia smithii, Sideritis canariensis, Smilax canariensis, Viburnum tinus.*

BIRDS YOU MIGHT SEE. **Sparrowhawk, Buzzard, Kestrel, Woodcock, Rock Dove, Bolle's Laurel Pigeon, White-tailed Laurel Pigeon, Turtle Dove, Long-eared Owl, Plain Swift, Berthelot's Pipit, Grey Wagtail, Robin, Blackbird, Sardinian Warbler, Blackcap, Chaffinch, Goldcrest, Blue Tit, Raven, Chiffchaff, Canary.**

BUTTERFLIES YOU MIGHT SEE. **Bath White, Brown Argus, Canary Blue, Canary Grayling, Canary Speckled Wood, Cardinal, Cleopatra, Clouded Yellow, Indian Red Admiral, Lulworth Skipper, Meadow Brown, Painted Lady, Red Admiral, Small Copper, Small White.**

Indian Red Admiral

95

PINE FOREST - *Pinar* (Excursions 13 - 16)

13. *Aguamansa*

Habitat description
Canary Pine forest at about 1200m (3900ft) with some areas including other tree species, and with varied undergrowth.

Special points of interest
Access to mature **Canary Pines** at El Topo.

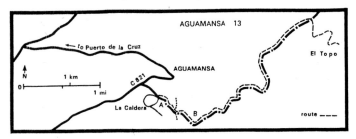

HOW TO GET THERE. *AGUAMANSA* is in the north of the island, at the east of the *OROTAVA* valley. **Bus:** See Appendix. **Car:** From the town of *OROTAVA* take the inland mountain road signposted to *"EL TEIDE"*. After about 9km (6mi) you will come to a left turn to *"LA CALDERA"*. This is about 2km (1mi) after the fish farm and arboretum, just above the village of *AGUAMANSA*.

PRACTICAL DETAILS. Easy tracks to walk on but the available maps for this area are quite confusing. If you intend to use this as a starting point for an extensive walk we would advise getting a copy of LANDSCAPES OF TENERIFE. You will need about 3hrs to walk to *EL TOPO* and back.

The excursion
The forests of Aguamansa lie on the eastern cliffs and slopes of the valley of Orotava. Whilst some of the forest is pine forest, much of it is mixed. Like the laurel forest on the cliffs at the west of the valley (see *CHANAJIGA*, EXCURSION No.12) the forest here is frequently shrouded in damp cloud for days on end. The area has many well marked walks and we shall describe just the way to *EL TOPO*. You could, in fact, drive the whole way.

Take the forest road past the bar at La Caldera car park and follow along the route shown in the sketch map, watching out for signs for *"LOS ORGANOS"*; it is a fairly level road for most of the way. Between Points

A and B the trees are mainly **Canary Pine**, some being very large and draped with grey lichens. The understory is of the *fayal-brezal* type (see ECO-LOGICAL ZONES, Section 1.) with the tree heath ***Brezo*** and **Myrica** dominating, along with **Laurus, Arbutus** and **Viburnum**. Among the smaller shrubs the most conspicuous (at least in early summer) are two species of **Cistus**: *C.symphytifolius* with large pink flowers and *C.monspeliensis* with narrow sticky leaves and smaller white flowers. On these you will sometimes find a brown spiny leaf beetle, *Dicladispa occator*, which is an endemic species that is very typical of the pine forest; its larvae feed only on **Cistus**. Other plants here include **Bramble**, a St. John's Wort, **Hypericum**, and also the white-flowered, red-berried **Daphne**. There are also several leguminous shrubs here; you will probably find **Chamaecytisus**, with white pea-flowers, and **Teline** and ***Codeso*** with yellow ones.

Beyond Point B there are more introduced trees, including **Sweet Chestnut** and a strong smelling **Eucalyptus**, while the undergrowth includes **Viburnum**, an **Echium** and **Rumex**.

After about 1hr walking you will reach a picnic shelter where you can turn right and start the climb up to El Topo. Here there are groups of ancient **Canary Pines** and you will also find **Chamaecytisus** and **Greenovia** in gullies. The **White-tailed Laurel Pigeon** occurs in this area, but you will have a better chance of seeing it in some of the laurel forest sites that we describe.

There is a rich array of invertebrate animals in this area. If you are lucky you may come across the well known **Praying Mantis** *(Mantis religiosa)* which disguises itself as part of a plant while waiting for its prey; it is, however, commoner in agricultural areas. Under rocks there is a chance of finding the small grey-brown wingless cricket *Gryllomorpha canariensis* and at least two endemic earwigs in the genus *Guanchia*. In the summer you are very likely to see one of the most spectacular insects of the island, a giant endemic asilid fly *Promachus vexator*, which is sometimes over 30mm long and is a voracious predator on other insects; a specially intense burst of buzzing usually indicates an attempt at mating.

SOME NOTABLE PLANTS. *Adenocarpus foliolosus, Arbutus canariensis, Castanea sativa* (**Sweet Chestnut**)*, Cedronella canariensis, Chamaecytisus proliferus, Cistus monspeliensis, C.symphytifolius, Daphne gnidium, Echium* sp.*, Erica arborea* (***Brezo***)*, Eucalyptus* sp.*, Greenovia* sp.*, Hypericum* sp.*, Ilex canariensis* (**Small-leaved Holly**)*, Laurus azorica, Myrica faya, Pinus canariensis* (**Canary Pine**)*, Rubus* sp. (**Bramble**)*, Rumex lunaria, Sideritis canariensis, Teline canariensis, Urtica morifolia, Viburnum tinus.*

BIRDS YOU MIGHT SEE. **Sparrowhawk, Buzzard, Kestrel, Barbary Partridge, Woodcock, Rock Dove, White-tailed Laurel Pigeon, Turtle Dove, Long-eared Owl, Berthelot's Pipit, Grey Wagtail, Robin, Blackbird, Sardinian Warbler, Blackcap, Chiffchaff, Goldcrest, Blue Tit, Chaffinch, Blue Chaffinch, Greenfinch, Canary, Linnet, Corn Bunting.**

BUTTERFLIES YOU MIGHT SEE. **Bath White, Canary Blue, Canary Speckled Wood, Cleopatra, Clouded Yellow, Large White, Lulworth Skipper, Meadow Brown, Small Copper, Small White.**

Asilid fly

14. *Montaña de Joco*

Habitat type
Fairly dense pine forest at about 1800m (5900ft) with shrub understory, on the north side of the dorsal ridge of the island.

Special points of interest
One of the few easily accessible areas with very old **Canary Pines**; also **Blue Chaffinches**, and rock faces with endemic succulent plants.

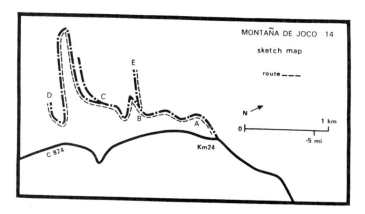

How to get there. *MONTAÑA DE JOCO* is about halfway along the mountain ridge that extends from *LAS CAÑADAS* to *LA LAGUNA*. **Bus**: Not possible. **Car**: on the road (C824) along the dorsal ridge from *LA LAGUNA* (near *SANTA CRUZ*) to *LAS CAÑADAS*. Just west of km post 24 there is a forest road leading off to the right, signposted *"PISTA AL CORTA FUEGO"*. You can park a short way down this road.

Practical details. This is an easy walk along a forest road with only slight gradients; no special footwear is necessary. You can drive along, but of course you will see less if you do this. There are numerous forest roads in this part of the island but there is no available map that shows them all. The accompanying sketch map should be sufficient for reaching the areas with the old pine trees, but if you wander elsewhere it would be wise to take a compass and the military map (which shows the contours) for this area, since the terrain can be very confusing especially when the clouds reach this level. Once you leave the road you will need a minimum of one hour to walk to the rocky point at the end of the track and return.

The excursion

Montaña de Joco is an area of **Canary Pine** covering a series of ridges and steep slopes at the head of Barranco del Pino, which cuts into the dorsal ridge of the island just east of Orotava. You can get a good view down this *barranco* from the road just before km post 25. From here you can see the areas that have been replanted, and also small areas of very old pines on some of the ridges. There is also a good view from *Mirador de la Cumbre*, which is 1km further along the road.

This forest area is less humid than the forests at lower elevations (see *AGUAMANSA*, Excursion No.13), although the cloud bank does frequently reach up to this level. The forest track winds in and out round the heads of several ravines, giving a splendid opportunity for easy access to some attractive rock faces with communities of the endemic succulent plants - **Aeonium** *(A.spathulatum)* and **Greenovia** - growing on them. You will also see a **Sideritis** and if you scramble up the gully at Point A you will find the beautiful yellow thistle **Carlina** *(C.salicifolia)*, and an endemic **Rumex** species that grows into a big bush.

Some of the ridges below on your right have the remains of ancient pine forest, although the higher slopes - where the track is - are planted. The shrub understory varies; in places it is mainly the tree heath *Brezo* and in other places it is predominantly either **Adenocarpus** or **Chamaecytisus** and some **Cistus** *(C.symphytifolius)*. The latter is one of the local shrubs which is not eaten by goats; it is not fire resistant but regeneration occurs rapidly from seeds which survive the passage of a forest fire. Along the side of the track you can find some **Bracken, Tolpis, Hypericum**, and a strongly scented **Bystropogon**. The populations of **Adenocarpus** here (and perhaps those of some other plants) show signs of past hybridization between species typical of the high mountain zone above and those found lower down: the opportunity for this may have arisen because of recent changes in the extent of the pine forest.

At Point B the track forks and goes up through an area of red pyroclasts (see GEOLOGY, Section 1). It soon forks again - at Point C. The left fork will take you up a fairly steep incline through an ugly (but necessary) fire break; after about 15 min you will reach a rocky outcrop with truly magnificent views and a few remaining old **Canary Pines** - Point D. If you take the right fork at Point B you descend a little and come out onto a ridge which has a crown of very old pine trees - Point E. These are lovely large spreading trees, with curtains of grey lichen *(Usnea* sp.) hanging down from their branches. Between the trunks you can get a good view of the island of La Palma and also of Mt.Teide and down into the valley of Orotava (or more probably onto a sea of cloud hiding the valley from you). There are **Blue**

Chaffinches here, but not so easy to see as in the pine forest picnic areas with tables and water drips. The conspicuous funnel webs along the sides of the track belong to the spider *Agelena canariensis.*

SOME NOTABLE PLANTS. *Adenocarpus* sp., *Aeonium spathulatum, Bystropogon* sp., *Carlina salicifolia, Chamaecytisus proliferus, Cistus symphytifolius, Erica arborea* (**Brezo**), *Greenovia aurea, Hypericum* spp., *Micromeria* sp., *Pinus canariensis* (**Canary Pine**), *Pteridium aquilinum* (**Bracken**), *Rumex lunaria, Sideritis* sp., *Tolpis* sp., *Usnea* sp.

BIRDS YOU MIGHT SEE. **Sparrowhawk, Buzzard, Kestrel, Barbary Partridge, Woodcock, Rock Dove, Turtle Dove, Plain Swift, Berthelot's Pipit, Robin, Blackbird, Spectacled Warbler, Chiffchaff, Goldcrest, Blue Tit, Blue Chaffinch, Canary.**

BUTTERFLIES YOU MIGHT SEE. **Bath White, Brown Argus, Canary Blue, Canary Grayling, Canary Speckled Wood, Cardinal, Clouded Yellow, Green-striped White, Indian Red Admiral, Painted Lady, Red Admiral, Small Copper, Small White.**

Blue Chaffinch

15. *Las Lajas*

Habitat description

Open **Canary Pine** forest with sparse undergrowth at almost 2100m (6900ft), close to the transition to the treeless high mountain zone.

Special points of interest

A good place to see many birds, especially the **Blue Chaffinch** and the Canary Island subspecies of the **Great Spotted Woodpecker**.

How to get there. *LAS LAJAS* is high on the southern slopes of the island, between *VILAFLOR* and *LAS CAÑADAS*. **Bus**: See Appendix. **Car**: Drive 5km (3mi) down from *BOCA TAUCE* at the southwest point of *LAS CAÑADAS* and turn off to the right .3km after km post 58.

Practical details. Easy walking on fine cinders. No special footwear needed unless you wander far.

The excursion

This is an area of high altitude forest in the dry south of the island; in fact Las Lajas is the highest forest picnic site on the island. You are not allowed to camp, but especially if you are here early in the morning or late in the evening you can watch **Turtle Doves, Blue Chaffinches, Great Spotted Woodpeckers, Canaries** and **Berthelot's Pipits** come and drink from the taps that are provided for the picnickers and are often left dripping; we have also seen **Hoopoes** and a bathing **Goldcrest**. This is not a densely planted commercial pine forest but an area with relatively dispersed **Canary Pines,** many of them quite old and growing very close to the tree-line (the level at which the pines peter out and the shrubs of the high mountain zone take over).

The area immediately around the picnic site is raked and kept tidy but if you walk away from here you can see the natural vegetation, which is is rather patchy. In some places there is a sparse undergrowth of **Chamaecytisus**, which is a very characteristic leguminous shrub of the pine forest, especially in dry parts of the island, and is one of the few shrubs that flourishes here among the dispersed pines high up on the southern slopes. It is tall and somewhat straggly, with small clusters of white flowers. In other places among the pines you will see more of a smaller yellow-flowered legume, *Codeso de la Cumbre*. Some rocky places have neither trees nor large shrubs, but only some of the small subalpine shrubs such as **Pterocephalus** (*P.lasiospermus*), **Scrophularia, Tolpis**, and a species of **Argyranthemum**. One interesting area is beyond the picnic site and football pitch where you can walk over areas of different coloured wind-borne volcanic debris (pyroclasts); these are in various shades of white (trachitic), red and greyish black (basaltic).

In an adjacent gully and further up towards the ridge the pines are absent and **Retama del Teide** takes over completely: this is the dominant broom of the treeless high mountain zone but it also grows in this transitional zone where the pines are unable to establish themselves in the less favourable patches.

There are two large Canary endemic beetles that have larvae which feed on dead pine trunks; one *(Buprestis bertheloti)* is black with a yellow pattern on its elytra (wing covers) and can often be seen flying, whilst the other *(Temnochila pini)* is brilliant blackish blue and can be found under the bark of dead pines. A conspicuous ground beetle is the endemic *Carabus interruptus* which is black with green highlights; you may also come across the very large slow moving black tenebrionid beetle *Pimelia radula.*

While you are in this part of the island it is well worth while driving down towards Vilaflor. Between km posts 63 and 66 the road passes through an area of pine forest that was severely burnt in spring 1984. The **Canary Pine** is one of the few species of pine that are fire resistant; the bark is charred but the trees themselves recover. A little further on at km 67 are two splendid examples of old **Canary Pines** close to the road.

SOME NOTABLE PLANTS. *Adenocarpus viscosus (**Codeso de la Cumbre**), Argyranthemum* sp., *Asphodelus* sp., *Chamaecytisus proliferus, Pinus canariensis* (**Canary Pine**)*, Pterocephalus lasiospermus, Scrophularia glabrata, Spartocytisus supranubius (**Retama del Teide**)*, *Tolpis webbii.*

BIRDS YOU MIGHT SEE. **Sparrowhawk, Buzzard, Kestrel, Barbary Partridge, Plain Swift, Rock Dove, Turtle Dove, Hoopoe, Great Spotted Woodpecker, Berthelot's Pipit, Grey Wagtail, Blackbird, Spectacled Warbler, Chiffchaff, Goldcrest, Great Grey Shrike, Raven, Blue Chaffinch, Canary.**

BUTTERFLIES YOU MIGHT SEE. **African Migrant, Bath White, Brown Argus, Canary Blue, Canary Grayling, Canary Speckled Wood, Clouded Yellow, Green-striped White, Painted Lady, Red Admiral, Small Copper, Small White.**

Canary Speckled Wood

16. *El Lagar*

Habitat description

Planted **Canary Pine** forest at 1000m (3300ft).

Special points of interest

Good place for visitors staying in the north of the island to see the **Blue Chaffinch** and several other island birds.

How to get there. *EL LAGAR* is on the north side of the island to the west of *PUERTO DE LA CRUZ* and south of *LA GUANCHA*. **Bus**: Only possible as far as *LA GUANCHA* - see Appendix. **Car**: In central *LA GUANCHA* there is a notice saying *"ZONA RECREATIVA EL LAGAR"*. Drive up this road and keep right where it forks very soon (unsignposted), up a road called *"CALLE LOS PINOS"*. Keep with this steep asphalt road to the extreme edge of the town and past the *casa forestal* (forestry office) on the left. 1.65km (1mi) after the *casa forestal* turn off the tarmac road to the right and up an unpaved road by a picnic shelter - the road is signposted *"EL LAGAR"*. Take a right turn after 1.9km (1mi) and another after a further 1.7km (1mi). You then come to a T-junction with a shelter and table. Go right here where it is well signed. In another 1km you arrive at the *zona recreativa*: it is 7km (4.5mi) from *LA GUANCHA*.

PRACTICAL DETAILS. Easy walking on forest road.

The excursion

El Lagar is an area of re-afforestation with fairly open undergrowth, mainly of **Cistus, Myrica** and *Brezo*. It is not so interesting to walk in as, for instance, parts of Aguamansa and Montaña de Joco (see EXCURSIONS No.13 and 14). However, it is a very good place to see the **Blue Chaffinch** in the same area as the striking local subspecies of the ordinary **Chaffinch.** You are likely also to see the **Great Spotted Woodpecker** and the very dark headed local **Blue Tit**; both these species can sometimes be seen pecking vigorously and tearing at the flaky bark of the **Canary Pines**, which harbours many insects and spiders.

Birds you might see. **Sparrowhawk, Buzzard, Kestrel, Barbary Partridge, Rock Dove, Turtle Dove, Long-eared Owl, Plain Swift, Great Spotted Woodpecker, Berthelot's Pipit, Robin, Blackbird, Chiffchaff, Goldcrest, Blue Tit, Chaffinch, Blue Chaffinch, Canary.**

Butterflies you might see. **Canary Blue, Canary Speckled Wood, Clouded Yellow, Green-striped White, Lulworth Skipper, Meadow Brown, Small Copper, Small White.**

HIGH MOUNTAIN ZONE
Teide National Park (Excursions 17 - 20)

The next four excursions are all within Teide National Park, which was established in 1954, and includes the *caldera* of Las Cañadas. (For a description of the high mountain zone encompassing the park see ECO-LOGICAL ZONES.) The park is administered by ICONA (*Instituto Nacional para la Conservación de la Naturaleza*) which has offices in La Laguna and Santa Cruz.

It is easy to visit the park by bus from Puerto de la Cruz. There are also coaches from most tourist resorts but the stops at any one place (apart from restaurants and the cable car terminal) are likely to be short. The distances in the park are quite large and it really is best to take a car. There are only two places where you can stay; one is the *Parador* - a government-run hotel - towards the west of the park, and the other is the *Refugio de Altavista* - a mountain refuge high on the slopes of El Teide. You should contact the Visitor Centre or the *Parador* if you wish to stay at the refuge; it is closed in winter, although some outbuildings may be left open.

On arriving in the park, it is worth spending some time at the *Centro de Visitantes* (Visitor Centre). It is open from 09.00 to 16.00 almost every day (but usually closed at lunch time). There is a small museum with displays covering the geological history of the park and its plants and animals; a visual presentation which lasts for 1/2 hr, with narration in several languages; and one of the best book shops (for natural history) that exists on the island, where you can of course get a map of the park. Particularly fine specimens of the **Canary Lizard** live around the building and water is left dripping to attract **Blue Tits, Canaries**, the **Canary Blue** butterfly and other local insects. You can obtain information here (or at ICONA head office) about guided walks in the park; these walks leave from both the Visitor Centre and the *Parador*.

17. *La Fortaleza*

Habitat type

High mountain habitat of ancient vegetated lava (typical of much of Las Cañadas) at about 2000m (6500ft); a *cañada*; the isolated cliff of La Fortaleza.

Special points of interest

Magnificent views of Teide and an area unspoilt by tourism; opportunities to see the most impressive species of **Echium** and many other endemic plants.

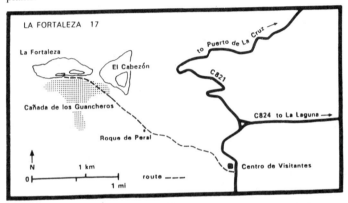

HOW TO GET THERE. *LA FORTALEZA* is in TEIDE NATIONAL PARK in the centre of the island. A footpath starts from behind the Visitor Centre. **Bus:** See Appendix. **Car:** From the north of the island drive up through *OROTAVA* on C821. It joins the C824, which runs along the mountain ridge from *LA LAGUNA*, near the east end of the *PARQUE NACIONAL DEL TEIDE*. The Visitor Centre is about 0.2km beyond this junction.

PRACTICAL DETAILS. Although you can walk as far as *LA FORTALEZA* unaccompanied, you are not allowed to climb the cliffs or go on any of the cliff paths without a ranger. It is best to check at the Visitor Centre before starting out on the walk; there are ranger-led walks most days. It is a good well-marked path, and no special footwear is necessary. Be prepared for considerable heat in the summer and take water, a hat, and sun cream; it is best to start early in the day and return before it gets too hot. In winter, be prepared for low temperatures. Do not pick any plants or disturb the rocks. You will need at least two hours for a leisurely walk to La Fortaleza and back. If you do not want to return the way you came you could follow Walk 13 in LANDSCAPES OF TENERIFE.

The excursion

We describe the visit to La Fortaleza in three parts. First there is the path, marked by small stones, which takes you through ancient well-weathered fields of lava and pumice (see GEOLOGY, Section 1.). Next there is the Cañada de los Guancheros, and then the cliffs of La Fortaleza.

On the slope immediately behind the Visitor Centre there are several well marked winding paths, ideal for short walks. Keep walking away from the building and you will eventually come to the path leading to La Fortaleza, with a view of Teide across the austere terrain to your left, probably with a cap of snow if you are there in winter. In the distance on your right you will see the tree covered mountain El Cabezón, which is at the extreme upper limit of the pine plantations. After about 20 minutes walking you will come to a sign saying *"ROQUE DE PERAL, CAÑADA DE LOS GUANCHEROS, EL CABEZON, LA FORTALEZA"*. In about another 20 minutes you will come to the isolated rock on your left known as Roque del Peral, and after a further 10 minutes you will reach the top of a steep slope above the Cañada de los Guancheros. The cliff wall of La Fortaleza dominates the other side of this low-lying sandy *cañada*.

In the spring and early summer this walk takes you through a blaze of colour which is enhanced by the sweet, almost sickly, scent of the flowers and the loud hum of insects. The ground is dotted with shrubs, with bare outcrops of dark lava and patches of pale pyroclasts (of the pumice type) in between. Nearly all the plants you will see are found only in the Canary Islands and all six of the commonest shrubs of Las Cañadas can be seen along this path. The most conspicuous plant is **Retama del Teide**, also called **Retama Blanca** (and sometimes just **Retama**, but this is really the name for another leguminous shrub with white flowers - *Retama raetam* - which does not occur so high up). This legume, which is closely related to broom, covers large areas of the high mountain zone; it forms tall dense shrubs and has white or sometimes pinkish flowers; its roots spread out a considerable distance from the plant. Almost as common is another leguminous shrub **Codeso de la Cumbre**, which is smaller, more like gorse (but not prickly) and with bright yellow flowers; it often grows in close association with **Retama del Teide**. There is also an **Argyranthemum,** growing in dense clumps, with large white daisy-like flowers. A very common crucifer - **Descurainia** - forms domed sheep-sized shrubs and has long spikes of yellow flowers. When the flowers of this plant die, the flower stalks remain and give it a very characteristic whitish appearance; during winter ice often accumulates on these spikes and then melts in the sun during the day and drains down towards the roots. There is another common but more straggly crucifer, **Erysimum**, which has mauvish pink flowers, and also a small cushion-like shrub scabious, **Pterocephalus**, with long-stemmed pink flowers. Of these six shrubs, four - the high mountain species of **Descurainia, Pterocephalus, Erysimum**, and **Argyranthemum** - occur only on Tenerife.

When the *Retama del Teide* is in flower, hundreds of beehives are brought up to Las Cañadas and the bee most in evidence then is the apiary variety *(Apis mellifera)*. However, there is also an endemic **bumble bee** - *Bombus canariensis* - which is black with a white-tipped abdomen. You might see an unusual endemic day-flying moth *Dipsosphecia vulcania* which belongs to a group of moths whose members are mimics of bees and wasps; it has transparent wings and a black body with pale stripes.

Another day-flying moth is the **Hummingbird Hawkmoth**, which may be seen here darting from blossom to blossom and hovering in front of them as it feeds. If you are observant and lucky you might see the mantid *Pseudoyersinia teydeana*, which has vestigial wings; it is a Tenerife endemic and is restricted to Las Cañadas. Two endemic weevils are common here; one is *Cionus variegatus*, which is pattened with creamy white and black blotches and feeds only on figworts *(Scrophularia* spp.); the other is *Cyphocleonus armitagei*, a large (2cm) species, black with white markings, which lives only on the shrub **Argyranthemum**. Another black and white insect is the bug *Eurydema ornatum*, which is sometimes very abundant on **Descurainia** plants. Among the butterflies you might see are the endemic **Canary Blue** and the **Green-striped White** which is not found lower on the island. This is only a selection of the interesting insects that you might see here; others are mentioned in Section 1, ECOLOGICAL ZONES.

Canary Lizards are very common throughout this area and form part of the diet of the **Kestrel**, several pairs of which breed on the cliffs of La Fortaleza; aggressive encounters between **Kestrels** and **Ravens** are quite a frequent sight. **Berthelot's Pipit** is also common here, and there is a good chance of seeing **Spectacled Warbler**, **Great Grey Shrike**, and coveys of **Barbary Partridge**. The endemic **Blue Chaffinch** does not breed in Las Cañadas, but apparently it sometimes makes trips up from the pine forest to feed in the bushes of *Retama del Teide*. **Rabbits** are common, sometimes living in holes in the lava rather than in burrows; they are hunted for sport during the autumn. Other mammals recorded from this area are the **Roof Rat, House Mouse, Algerian Hedgehog** and bats, especially the **Canary Long-eared Bat**. Both feral **Cats** and **Dogs** survive in this area, but the **Cats** are rarely seen.

When you descend onto the flat Cañada de los Guancheros, you will be immediately aware of the lack of vegetation, except for a few *Retama del*

Teide bushes at the edge and the occasional seedling that germinates in the pumice sand and then dies. This is a very inhospitable environment for most plants and with extreme climatic conditions (see GEOLOGY, Section 1 for discussion of *cañadas*). In the hot summer the sand, composed of granular and dusty pumice, dries out whilst after heavy rain or snow the whole area may become waterlogged for several days. On winter nights the *cañada* forms a frost pocket, holding cold air that flows down from the surrounding areas. The sand particles here are often actually moved by frost action: this happens when the water in wet sand freezes and the ice expands forming small ridges or bumps on the surface; as the ice melts in the daytime sun, sand grains slip down from these bumps into the dips in between. As this process happens day after day, sand grains of different sizes accumulate at slightly different places; the patterns resulting from this 'frost heave' can be seen on the surface of the sand.

In spite of the severe conditions in the *cañada*, some invertebrate animals manage to live here. The most easily seen are small jumping spiders belonging to a species (*Phlegra lucasi*) which can also be found on lava fields totally lacking vegetation. Under the few rocks that are scattered over the surface you can find a wider variety of animals: there are beetles, centipedes, small cockroaches (*Phyllodromica bivittata*), bristletails, and several species of spiders belonging to the family Gnaphosidae. This is a group that is very successful in dry places, and members of many species live in silken sacs when resting by day; at night they hunt their prey rather than spinning snares. Most of these *cañada* invertebrate animals are nocturnal and avoid the heat of the day by sheltering under rocks which are large enough to remain relatively cool underneath.

After crossing the *cañada* you come to the cliffs of La Fortaleza. While the wall of the *caldera* of Las Cañadas is more or less complete in the south, all that remains in the north is this isolated cliff of La Fortaleza and the mountain of El Cabezón. These cliffs and the slopes below are a refuge for some very rare plants, a few of which grow here and nowhere else. You are not allowed to climb the cliff or the slopes below but in early summer you will be able to see the spectacular red flowers of an endemic **Echium** *(E.wildpretii)*; this is a close relative of Vipers Bugloss and is the most famous flower of Tenerife, with flowering stems growing several metres high. Also conspicuous on the slopes is the giant yellow-flowered **Ferula**, and high on the cliff face some **Juniperus** (*J.cedrus*). The latter is one of the plants that is making a conspicuous recovery from the heavy grazing of goats and cattle, which was outlawed in this area about 40 years ago. A large white-flowered leguminous bush growing here is **Chamaecytisus**, a plant found in the understory of parts of the pine forest, but which forms the dominant vegetation in some southern areas high on the outer slopes leading up to the rim of the *caldera*. You will also see the plantain **Plantago**, and an attractive **Sideritis**, which is a sage-like plant with white woolly leaves and spikes of pale yellow flowers. A rare yellow thistle, **Carlina** *(C.xeranthemoides)*, grows profusely here, and you will also find a pink-flowered species of **Cistus** *(C.osbaeckiaefolius)* which occurs only here at La Fortaleza, and also an attractive tall grass *(Arrhenatherum calderae)* which is almost as restricted in its range. Two other Canary endemic plants that can be seen here are the pink-flowered campion **Silene** *(S.nocteolens)* and an elegant **Tolpis** - a composite with slender branched stems bearing small brilliant yellow flowers.

SOME NOTABLE PLANTS. **On the walk**: *Adenocarpus viscosus (***Codeso de la Cumbre***), Andryala pinnatifida, Argyranthemum teneriffae, Bystropogon origanifolius, Descurainia hourgeauana, Echium auberianum, Erysimum scoparium, Nepeta teydea, Pterocephalus lasiospermus, Scrophularia glabrata, Spartocytisus supranubius (***Retama del Teide***), Tolpis webbii*. **At La Fortaleza**: *Aeonium spathulatum, Arrhenatherum calderae, Carlina xeranthemoides, Chamaecytisus proliferus, Cheirolophus teydis, Cistus osbaeckiaefolius, Echium wildpretii, Ferula linkii, Juniperus cedrus, Lotus campylocladus, Micromeria* sp., *Monanthes niphophila, Pimpinella cumbrae, Plantago webbii, Rhamnus integrifolia, Sideritis* sp., *Silene nocteolens, Tolpis webbii*.

BIRDS YOU MIGHT SEE. **Kestrel, Barbary Partridge, Rock Dove, Plain Swift, Berthelot's Pipit, Spectacled Warbler, Chiffchaff, Great Grey Shrike, Raven. Near the houses at El Portillo: Blackbird, Blue Tit. Non-breeding visitors: Long-eared Owl, Hoopoe, Robin, Blue Chaffinch, Canary.**

BUTTERFLIES YOU MIGHT SEE. **Bath White, Canary Blue, Canary Grayling, Green-striped White, Painted Lady, Small Copper** (occasional)**, Small White.**

Cistus osbaeckiaefolius

18. *Pico del Teide* (Peak of Teide)

Habitat type
Unvegetated volcano peak in the high mountain zone.

Special points of interest
Hot air vents (fumaroles) with sulphur deposits, high volcanic crater 3718m (12198ft) and superb views.

HOW TO GET THERE. *PICO DEL TEIDE* is in the *TEIDE NATIONAL PARK* in the centre of the island. You can either go up by *teleférico* (cable car) or walk up from *MONTAÑA BLANCA* or *LA FORTALEZA*. Bus: See Appendix. Car: Drive through the park from the eastern entrance boundary for about 7km (4mi) for *MONTAÑA BLANCA* track and about 9km (6mi) for *teleférico*. There is very limited parking space at the bottom of the *MONTAÑA BLANCA* track.

PRACTICAL DETAILS. The *teleférico* takes you near the top but there is still another 170m (560ft) to climb. It can be cool here in summer, and very cold in winter. Be prepared for wind, although the cable car doesn't run when it is very windy. Sun protection and lip cream are advisable. The *teleférico* trip takes about 8 minutes but you may have to stand in a queue for over an hour; it is best to get there at about 09.00. There are three trails starting at the upper *teleférico* terminal; one leads to the top of Teide and the others are more on the level and lead to spectacular viewpoints overlooking Pico Viejo and La Fortaleza. The time it takes to walk to the top of the volcano is about 20min but depends on how easy you find it to climb at this high altitude. Visitors are not allowed to wander from the marked paths. If you want to climb up from the Montaña Blanca side, and back down in the same day, plan it carefully. Perhaps the most interesting way to make the ascent is to stay the night at *El Refugio de Altavista* and climb the peak in time for sunrise on the following day (inquire as above). The track is well marked - see Walk 9 in LANDSCAPES OF TENERIFE; you will need about 3hrs for the ascent from the road to the top.

The excursion
Teide is one of the major volcanoes of the world, in the same class as Popocatepetl in Mexico, Kilimanjaro in Kenya and Vesuvius and Etna in Italy. It is dormant; not extinct - just sleeping. It is a strato-volcano - a volcano resulting from many eruptions, each of which buried the previous ones - and is made up of both pyroclasts (cinders and ashes that are thrown into the air by explosions) and lava (which pours out of the ground and flows down the mountain). Although there is often snow on Teide in the winter and small patches can sometimes be found in sheltered situations as late as June, it does not have snow all the year round, as sometimes stated in tourist literature; the whitish rock near the summit has obviously deceived some people into thinking that it does.

El Pico del Teide is the top of Teide and has a more extreme climate than the rest of the high mountain zone; almost no flowering plants grow above about 3200m (10500ft) except for the **Teide Violet** which is found up to about 3500m (11500ft). Whether you visit the peak under your own steam or go up by *teleférico* it will be an exciting experience. Although there has been no eruption from the top of this volcano in historic times, there was an eruption on the slopes of Pico Viejo (just to the west of Teide) in 1798.

If you climb up from the Montaña Blanca side you pass from an area of patchy shrubs to an area almost devoid of plants. On the lower slopes you will find *Retama del Teide, Codeso de la Cumbre*, **Descurainia** and **Argyranthemum.** After you pass the mountain refuge *(Refugio del Altavista)* at 3270m (10730ft) you will be walking on a large area of "aa" lava where no plants grow, although *Retama del Teide* reaches the lower edge of this flow. The **Teide Violet** grows elsewhere at this altitude where the ground is suitable.

If you go up by *teleférico* you will get a bird's eye view of the changes in vegetation. Around the upper terminal you may find an occasional **Argyranthemum** plant or a few grasses (for instance *Poa annua* and *Vulpia bromoides*) that were probably helped up here by humans. However, the **Teide Violet** is native at this level and a cudweed (*Gnaphalium* sp.) also manages to grow in a few places very close to the summit. As you climb to the top you will see small vents (fumaroles) with sulphur encrustations around them and sulphurous steam coming out of them; there is usually a strong smell of sulphur. A few green algae and mosses grow near these vents.

As you approach the top you will see that there is a circular crater. You are no longer allowed to walk down into it because it has become very worn from the trampling of so many tourists. From the crater rim, on a clear day, you can see the mountain tops on the islands of Hierro, La Palma, Gomera and Gran Canaria; the lower parts of these islands are often obscured by a sea of clouds (see CLIMATE, Section 1.). More often than not, much of the north side of Tenerife itself is also obscured by a huge mantle of cloud. When basking in the sun up here one can search in vain for Puerto de La Cruz, which will be having yet another overcast day.

If you are up here in winter and look carefully at the surface of the snow, you are likely to see the dead bodies of a wide variety of insects, and perhaps some that are still alive, though chilled and sluggish. Insects may also be quite conspicuous near the peak on summer days, either in flight or settled on the rocks. The smaller ones, such as aphids, evidently reach the mountain top by accident, since their normal dispersal flights involve relatively passive drifting with the wind, and landing almost at random. Some stronger-flying insects seem to be attracted by high places, and can often be seen and heard buzzing in front of the rock faces. One striking example on Teide is a large black hoverfly *(Scaeva pyrastri)*, which has a double row of bright yellow marks on its abdomen. Many of the insects that reach the mountain top eventually die there, and we have found that there are a number of kinds of invertebrates that manage to make a living at this high altitude by scavenging on the dead and dying stranded insects. For instance, the large harvestman *Bunochelis spinifera* (a long-legged spider relative) can often be found in shady places on the rim of the summit crater. Another scavenger that we have found only a few metres below the peak is a centipede *(Lithobius crassipes)*, while close to some fumaroles at 3500m (11500ft) on the path down to Montaña Blanca there is an abundant population of a bristletail *(Ctenolepisma longicaudata)*; a black tenebrionid beetle *(Hegeter lateralis)* can also be seen here.

Vertebrates are scarce at these high altitudes, but one can sometimes see **Canary Lizard**s near the summit crater, and the ubiquitous **Berthelot's Pipit** also occasionally visits the peak. One recent visitor was lucky enough to see a group of Choughs *(Pyrrhocorax pyrrhocorax)* soaring and calling around the peak. Choughs do not breed on Tenerife, and are very rarely seen there; these birds must have been stragglers from the isolated population on La Palma, whose closest relatives are in northwest Africa.

SOME NOTABLE PLANTS. *Adenocarpus viscosus* (**Codeso de la Cumbre**), *Argyranthemum teneriffae, Descurainia bourgeauana, Gnaphalium* sp., *Poa annua, Spartocytisus supranubius* (**Retama del Teide**), *Viola cheiranthifolia* (**Teide Violet**), *Vulpia bromoides.*

BIRDS YOU MIGHT SEE (not breeding). **Kestrel, Rock Dove, Plain Swift, Berthelot's Pipit, Raven.**

BUTTERFLIES YOU MIGHT SEE. **Canary Blue, Green-striped White, Painted Lady.**

Teide Violet

19. *Los Roques de García*

Habitat description

Eroded remnants of an ancient crater wall at about 2200m (7000ft), with adjacent old vegetated lava flows.

Special points of interest

High mountain vegetation; view down into the largest *cañada*; dramatic rock formations separating upper and lower segments of the *caldera*; "pahoehoe" and "aa" lava.

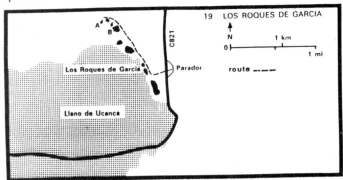

HOW TO GET THERE. *LOS ROQUES DE GARCIA* are in *TEIDE NATIONAL PARK* in the centre of the island. **Bus**: See Appendix. **Car**: Drive 13km (8mi) from the eastern end of the park as far as the entrance to the *Parador*. Opposite (on the right of the road) is a turning leading to a parking place.

PRACTICAL DETAILS. Although this is a relatively easy walk, stout shoes are recommended, since some of the way is over fairly uneven ground. You will need 1hr to walk to the furthest of the rocks and back. There are ranger-led walks here; details from the *Centro de Visitantes* (Visitor Centre) or the *Parador*.

The excursion

Los Roques de García are apparently the remains of a rim between the two major parts of the *caldera* of Las Cañadas (see GEOLOGY, Section 1). They are a row of spectacularly eroded rocks stretching from the southern wall of Las Cañadas to the lower slopes of Teide. The weird shapes of some of them are due to differential erosion. The upper parts are lava which is more resistant to erosion whilst the lower parts are composed of less resistant layers of compressed pyroclasts.

From the parking place you can look down into the Llano de Ucanca, which is part of the remains of a smaller southwestern crater about 70m (230ft) lower than the main crater. The floor of the lower crater is now covered with fine volcanic sand which has been produced by thousands of years of erosion from the surrounding mountains. You can see how the tips of the tongues of dark lava, which flowed down the slope on your far right, are now submerged under accumulated sand. When the snow melts in the spring, the whole of the low area can become a lake for several days. There are some patches of *Retama del Teide* and **Pterocephalus** but most of the *cañada* is bare.

Below you on the rubble slopes of Los Roques grows a tall red-flowered **Echium** *(E.wildpretii)* together with some small **Retama del Teide** bushes, **Pterocephalus** and **Plantago**. If you walk up the steps and up the nearby rock you will get a magnificent view across to Guajara - which at 2715m (8907ft) is the highest mountain in the southern rim of Las Cañadas - and a bird's eye view of much of Las Cañadas and its vegetation. The dominant plant between you and Guajara is *Retama del Teide*; it even grows on the face of the mountain wherever it can get a root hold. Between you and the *parador* there are also large clumps of the pink scabious **Pterocephalus**, contrasting in early summer with the white flowered *Retama del Teide*. Until about 40 years ago **Pterocephalus** was rare - presumably because of goat and cattle grazing; it has now made a spectacular recovery. Near the road you can see **Descurainia**, which for most of the year has straw-coloured dead flower stems which give it its characteristic cushion-like look; it has brilliant yellow flowers in early summer. Less abundant here but also forming hummocks is *Codeso de la Cumbre*, again with yellow flowers. Immediately around you, on this rock, grow **Tolpis, Scrophularia, Argyranthemum** and **Lotus**.

If you now descend the few steps from the parking lot and walk along the broad sandy track by Los Roques you will find that *Codeso de la Cumbre*, **Descurainia** and a few *Retama del Teide* dominate the way. The track goes through a picnic site and an area where you will see the stumps of a large number of felled trees. This area was planted with a non-native species of pine, but the more enlightened conservation policy of recent years has resulted in these trees being cut down as part of an attempt to keep the park as close to its natural state as possible. From here on the path is not well marked, but head for the last (northernmost) of the rocks. You will pass through a fairly well vegetated gully, where you will almost certainly see **Great Grey Shrike** and the ubiquitous **Canary Blue** butterfly, as well as many of the other insects described in the visit to La Fortaleza (see EXCURSION No.17). As you approach the end of Los Roques you come to an area of "pahoehoe" lava which contrasts with the nearby "aa" lava. This is a good opportunity to compare these two main types of lava: the "aa" is chaotic and sharp while the "pahoehoe" is much smoother, either fairly flat on the surface or ropy. If you continue along, with the last of Los Roques de García on your left, you will find a beautiful look-out point over the Llano de Ucanca. You can see where the nearest part of the "pahoehoe" flow stopped right at the edge of the cliff (point A on map); in contrast about 75m (80yds) further back the lava went over and spread out below (point B on map).

SOME NOTABLE PLANTS. *Adenocarpus viscosus* (**Codeso de la Cumbre**), *Andryala pinnatifida, Argyranthemum teneriffae, Bystropogon origanifolius, Descurainia bourgeauana, Echium auberianum, Echium wildpretii, Erysimum scoparium, Nepeta teydea, Pimpinella cumbrae, Plantago webbii, Pterocephalus lasiospermus, Scrophularia glabrata, Spartocytisus supranubius* (**Retama del Teide**), *Tolpis webbii.*

BIRDS YOU MIGHT SEE. **Kestrel, Barbary Partridge, Rock Dove, Plain Swift, Berthelot's Pipit, Spectacled Warbler, Chiffchaff, Blue Tit, Great Grey Shrike, Raven, Canary.**

BUTTERFLIES YOU MIGHT SEE. **Bath White, Canary Blue, Canary Grayling, Green-striped White, Painted Lady, Small Copper** (occasional), **Small White.**

Echium wildpretii

20. *Las Narices del Teide* - Teide's Nostrils

Habitat description
High altitude lava flow at 2000m (6550ft), in the high mountain zone.

Special points of interest
One of the most recent lava flows on the island (1798) - easily accessible by car; many features of volcanic topography.

How to get there. *LAS NARICES DEL TEIDE* is in *TEIDE NATIONAL PARK* in the centre of the island. **Bus**: See Appendix. **Car**: Take the C821 across *LAS CAÑADAS* as far as the road junction - *BOCA DE TAUCE* - in the southwest end of this vast crater. One road leaves the crater at this point. If you drive along the other road (C823) it will take you northwest in a straight line across a large expanse of lava. After about 1km you will find that there are two or three places where you can pull off the road. Take care, because it is easy to get stuck in a patch of cinders. It is best to stop at the first possible pull-off place, because from here you can descend into the low area of lava on your right.

PRACTICAL DETAILS. You will need sturdy shoes if you walk down off the road: on the recent unweathered lava it is best to have tough boots and a pair of gloves to protect your hands. Be prepared for great heat during the day in summer and sub-freezing temperatures in early morning and evening in the winter.

The excursion
Las Narices del Teide - Teide's Nostrils - is the name given to the twin vents high on the slopes of Pico Viejo (the old peak) and the lava flow that poured out in 1798. This is the most recent flow in Las Cañadas and one of only about half a dozen flows on the island that have occurred in historical times. The latest of these was Chinyero in 1909: it can be reached by driving another 7km (4mi) further along the C823.

You may not want to walk on the lava, but many features of this area can be appreciated from the roadside. Assuming that you are about 1km from Boca de Tauce, you will be able to see where the lava flowed down in a broad wandering river and reached the wall of Las Cañadas just to the west of where you are standing. The paler areas visible high up in this flow are called "kipukas" - a word which comes from Hawaii - or *islotes*: they are small islands of earlier lava which were not covered by the most recent flow. The contrast in the vegetation is immediately obvious. The recent unweathered dark "aa" lava - *malpaís* - has virtually no vegetation, while the paler, older, adjacent lava has sparse shrubs and smaller plants. This is because 200 years is not enough for lava at this altitude and in this climate to become weathered and accumulate enough debris to form soil for plants. The adjacent lava is very much older.

The road at this point is on part of the 1798 flow (somewhat disturbed by road building). If you climb down from it you will reach an area of older "pahoehoe" lava. Here you will find ***Retama del Teide, Codeso de la Cumbre*, Argyranthemum, Scrophularia, Pterocephalus** and a few patches of moss and lichen in cracks. This is one of the places where you might get a glimpse of a few **Mouflon** (introduced wild sheep) grazing in the distance. **Rabbits** are common here and you will see droppings which are deposited in piles and are very slow to disintegrate in this climate. **Canary Lizards** are also common; these are preyed upon by the **Kestrels** which you will almost certainly see hovering overhead. In summer the most conspicuous insect around here is the butterfly **Canary Blue**, but one can also see ant-lions *(Myrmeleon alternans)* - members of a group that is well represented in the Canaries but much less so in northwest Europe. Ant-lion adults are rather like dragonflies - with two pairs of large wings and long thin bodies - but their flight is much weaker and they have conspicuous stout antennae.

At the far side of this low area of "pahoehoe" lava there is "aa" lava of much the same brown colour with scattered vegetation; this type of lava continues up the mountainside. The left (as you look up the mountain) of the low area of "pahoehoe" lava is overhung by a wall of lava formed by the edge of the 1798 flow. This is unweathered and blackish "aa" lava, with no vegetation apart from the very occasional plant that has gained a foothold near the edge; it is extraordinarily sharp and unstable and is rather dangerous to walk on. You would not find the **Canary Lizard** or **Rabbit** here. On first glance it would seem improbable that anything lives on it at all. However, we have shown during our research on the island, that this type of recent unvegetated lava does support an interesting community of invertebrates, principally bristletails, crickets, centipedes and the earwig *Anataelia canariensis*. Some of these are endemic species and are more or less restricted to such spartan habitats; a few are almost absent from the more vegetated older lava a few metres away. This community of small animals is unusual in that it does not depend upon plants as the basis of the food chain, but exploits the 'fallout' of aerially dispersing and drifting insects; its members are carnivores and scavengers, rather than herbivores.

SOME NOTABLE PLANTS. *Adenocarpus viscosus (**Codeso de la Cumbre**), Argyranthemum teneriffae, Pterocephalus lasiospermum, Scrophularia glabrata, Spartocytisus supranubius (**Retama del Teide**).*

BIRDS YOU MIGHT SEE. **Kestrel, Barbary Partridge, Rock Dove, Plain Swift, Berthelot's Pipit, Chiffchaff, Raven, Great Grey Shrike.**

BUTTERFLIES YOU MIGHT SEE. **Bath White, Canary Blue, Canary Grayling, Green-striped White, Painted Lady, Small White.**

OTHER HABITATS (Excursions 21 - 24)

21. *Cueva de San Marcos* (St. Mark's Cave)

Habitat type
Volcanic lava tube, about 1.8km (1.1mi) long, in the cultivated lower zone of the north coast of the island, opening part way up a sea cliff.

Special points of interest
An underground walk with a chance to see small troglobites (animals adapted to living in the complete darkness of caves).

How to get there. *CUEVA DE SAN MARCOS* is near the coast close to *ICOD DE LOS VINOS* which is midway between *PUERTO DE LA CRUZ* and *BUENAVISTA*. **Bus**: See Appendix. **Car**: Coming from the east on Route 820 drive into ICOD. Leave the 820 on a right fork labelled *"PLAYA SAN MARCOS"*. Drive 0.7km down this road, ignoring one small road on the left and two on the right, until you come to roads entering from the right and left at the same point. Two striking large white houses on pillars, with red tiled roofs (Nos 12 and 14) are on your right. Turn left and go down this small road with walls on either side, for 200m (220yds). (You will see a big red-tiled house with a tall thin cyprus at the end of the road.) Ignore the first right turn. Ignore the next entrance before a low red-roofed building but stop at the entrance immediately after it. There are two green metal posts for a chain across this entrance. Set back from the road is another low red-tiled house and a single tall palm tree along a track. Walk down this track; at about 65m (71yds) beyond the solitary palm tree on your left, up over some loose rocks, is a group of palm trees. There are two small terraces more or less enclosed by vegetation. The cave entrance is at the side of the upper terrace (probably with Bananas): it is close to an orange tree and some other fruit trees, and is likely to be partially obscured with brush wood.

Practical details. This is not a trip for small children or for anyone who does not like the dark, or somewhat confined spaces. Two good torches and spare batteries are essential: it is totally dark in the cave. Although you can walk upright for most of the way, there are a couple of places where it is necessary to stoop, so some protection for your head is a good idea; the rock is very sharp. It is sometimes quite wet underfoot. Allow about one hour to reach the cliff face and return.

The excursion
The Cueva de San Marcos is a lava tube. These tubes occur in an area of "pahoehoe" lava and are formed when rapidly flowing lava carves out a channel to flow in. Even when the surface lava has solidified, liquid lava continues to run underground until finally, when the flow ceases, a hollow tube is left. This type of cave is generally very close to the surface, and quite

often one can stamp on the "pahoehoe" lava and hear the dull echo from a space beneath.

There are several lava tubes on the island but this is one of the most accessible and also one of the longest. It is actually part of one of the most complicated networks of lava caves that has been found anywhere in the world; much of it is still unexplored. The Cueva del Viento, the other side of Icód, has 15km (9.3mi) of mapped passages and ranks as the longest volcanic cave complex in the world (see CAVES, Section 1). Cueva de San Marcos is a mere 1.8km (1.1mi) long, but that is impressive enough for someone who is not used to walking underground.

The entrance to the cave is fairly low, but once you are through this you can stand up in a small chamber. The cave has been well explored and we can assure you that this is the only known entrance (apart from the one on the cliff face); the cave does not join up with any of the other caves - at least not via any passage that a human could use. On your left is an easy way down into a spacious part of the lava tube. The entrance is part way along the lava tube so you have a choice of whether to go to the right and along the tube as far as the cliff above Playa de San Marcos or through a small entrance to the left (with a white downward pointing arrow above it) and explore along the part of the tube that continues inland.

If you take the right hand route first you can walk upright most of the way and the floor is fairly smooth due to the accumulation of water-borne sand. The tube is high and broad where it is fairly level, but gets narrower where it slopes downward more steeply. Features such as the broad ledges running lengthways along the sides of the tube bear witness to the various stages in the formation of the tube. Notice small lava 'stalactites', sometimes encased in minerals from penetrating water. It is easy walking here because there is a considerable accumulation of soil on the floor brought in by water seeping from the banana plantation above. It is very humid but can be quite draughty in this section. You will see "light at the end of the tunnel" after 250m where the cave opens high on the cliff above Playa de San Marcos. Notice the water channel here. The plantation owner has used this section of the cave as a natural water pipe, and water is channeled from the cave exit into a water tank. When you look out over the little bay you can see a **Dragon Tree** on the cliff opposite; there are not many left growing wild.

If you return and explore the inland route (following the white arrow) you need to crouch through a low entrance at first but then come immediately into a large chamber. Sometimes this chamber has a lake in it, when the farmer has sent water down. The tube branches. If you take the right branch you will find the going rough, with raw lava underfoot; soil has not accumulated here. There is no exit at the end of this branch, so there is no air getting in to make a draught.

Since no plants grow in the cave, the animals that live here depend for their food on debris washed in by rain water, and in some cases on the roots of plants that grow on the surface above; others, of course, are predators. Bones of the extinct **Giant Rat** have been found here and also those of the extinct **Giant Lizard**. Occasionally bones of the **Canary Lizard** and **Rabbit** may be found, presumably from individuals that have wandered in and got lost; they do not live down here.

Twenty one species of invertebrates have been found in the cave, but the occurrence of some of these is accidental, and a number of others are found only near the entrance in the region of dim light. A third group are evidently successful members of the cave community, but are not specialized for cave life and also occur in surface habitats. They include the predatory centipede *Lithobius pilicornis*, two kinds of millipedes and a woodlouse, *Haplophthalmus danicus*; the latter three species all depend on decaying organic material. The woodlouse is of special interest since it has previously been recorded in the litter of the laurel forests of the Anaga peninsula. In addition, however, there are two true troglobites - animals that are found only in complete darkness. One is a white eyeless spider *(Lepthyphantes oromii)* and the other a blind cockroach *(Loboptera subterranea)*; both have only recently been discovered and described.

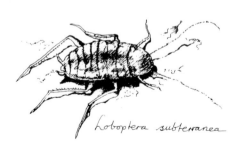

Loboptera subterranea

22. *Punta Gotera*

Habitat description

Ancient sand dune, buried by lava flows but now exposed by marine erosion. This dune is known locally as *MANCHA DE LA LAJA*.

Special points of interest

Marine and freshwater fossil shells.

HOW TO GET THERE. *PUNTA GOTERA* is close to the Nautical Club to the west of *BAJAMAR* on the north coast of the island. **Bus:** See Appendix. **Car:** Drive along the road from *TEJINA* towards *BAJAMAR* and take a small unsignposted turning to the left before km post 58. The turning is beyond a Banana plantation on the right, and on a right hand bend - it is further from *BAJAMAR* than would appear from the *Cabildo* map. There is a small planted triangle at the beginning of the side road and then it bends to the left past some 7-story buildings and goes straight down to the *Club Náutico*. If you stop by house No.37 on the left before reaching the club you can look down towards the shore and get a view of part of the buried dune. You can then make your way along the top of the cliff and down a track on the other side.

PRACTICAL DETAILS. An easy walk. Please don't deface the sandy cliff or remove fossils. A five minute walk from house No.37.

The excursion

Punta Gotera (or Mancha de la Laja) has seaward facing cliffs which expose a cross section of an ancient dune of pale shell sand of marine origin. The sand is about 20m (66ft) above the present level of the sea and up to 8m (26ft) deep. The dune was probably formed around 100,000 years ago and at that time must have been close to sea level: the land has evidently risen - or the sea has gone down - since that time. At some stage after its formation the dune was covered by massive lava flows, and it lies on a yet older shelf of lava. There are now no natural shell sand beaches along the coast of Tenerife - only beaches of volcanic sand. The calcareous sand from this site has been extracted commercially until relatively recently: this is no longer allowed because of the danger of undermining the land above. Much of the ancient dune structure is visible on the exposed cliff faces, and thin layers of successive depositions can easily be seen. Towards the top of the dune pyroclastic volcanic material is mixed with the calcareous sand and the orange coloured layers indicate that soil had started to form. The reddish bands adjacent to the dark lava are where this young soil was scorched by the molten lava. The most western exposure clearly shows several successive layers of lava above the dune. Some parts of this ancient dune abound with fossil mollusc shells, those of marine origin being at a lower level than those of terrestrial origin. Fossil bones of birds and of the extinct **Giant Lizard** have also been found here.

In the loose sand at the bottom of the cliff you may find small conical pits, a few centimetres in diameter. These are made by larvae of a species of ant-lion *(Myrmeleon alternans)*; these are predatory insects related to the lacewings. The larvae excavate a pit and bury themselves just below the surface of the sand at the bottom. When an insect, such as an ant, slithers down the side of the pit, the larva reaches out and grabs it in a flash with its enormous jaws. Try feeding one.

This small area is of considerable scientific interest and members of ATAN (a local conservation group - *Asociación Tinerfeña de Amigos de la Naturaleza*) are trying to ensure that steps are taken to protect it; meanwhile they are organizing voluntary conservation work here.

Fairly near Punta Gotera is one of the few artificial freshwater pools or settlement ponds with a soil and rock - rather than concrete - bottom. (This is about 1km back along the road to Tejina, on the seaward side of the road.) It is probably the only place on the island where you will see **Moorhens**. If you turn over some rocks around the edge of the pond you may find a shiny lizard with very small legs. This is the **Canary Skink**, which is quite common on the island but much less conspicuous than the **Canary Lizard** because it normally remains hidden by day. This is also a good habitat for the **Stripeless Tree Frog** and **Marsh Frog**. You can readily find a large dytiscid beetle, *Cybister tripunctatus*, swimming in the water; this beetle dives deep into the water but frequently comes to the surface to breathe.

High on the slopes above Bajamar is one of the few places where **Canary Palms** can be seen growing in a more or less wild state.

Fossil dune at Punta Gotera

23. *Punta del Hidalgo*

Habitat description

Intertidal zone with large expanse of rocky shelf exposed at low tide giving access to numerous rock pools.

How to get there. *PUNTA DEL HIDALGO* is almost the most northerly point on the island, about 7km (4.5mi) northeast of *TEJINA*. **Bus**: See Appendix. **Car**: Drive eastwards along the road above the shore between *BAJAMAR* and *PUNTA DEL HIDALGO*; you will see a high-rise building below on the shoreline. Keep on this road until you see a side road to the left opposite a small supermarket and pharmacy (before *PUNTA DEL HIDALGO* town is marked on the *Cabildo* map). Go straight down this side road and when it reaches the high rise building turn right along the shore road and continue for about .65km, where you will find easy access to the shore.

Practical details. Volcanic rock is very uncomfortable to walk on; you will need shoes that you don't mind getting wet. Snorkel and mask will increase your enjoyment, especially at high tide, but the sea is often rough on this side of the island.

The excursion

Punta del Hidalgo is a promontory of land in the extreme north of the island. It has about 2km of shoreline, consisting of a broad rocky shelf which is exposed at low tide (see OCEANOGRAPHY, Section 1.). It is easy to walk out here and explore the numerous rock pools. Until about a decade ago this area was very rich in marine life, but now it is suffering from the effects of too much beach camping, overfishing and some oil pollution. You can still, however, find a great variety of life here in the intertidal zone. One reminder - do always replace rocks that you disturb: the small animals that live below rocks can die very quickly when exposed to the air and sun or to predatory animals. Just leave things as you find them.

There is not a great variety of life in the higher rock pools, which are, of course, exposed for longer each day than those lower down. You will probably find a shrimp *(Palaemon elegans)* and two species of fish - a very small goby *(Mauligobius maderensis)* and a blenny *(Blennius parvicornis)*. The rocks round about are covered with barnacles *(Chthamalus stellatus)* and there are also limpets *(Patella piperata)*, winkles *(Littorina striata)* top-shells *(Osilinus atratus)* and a whelk *(Thais haemastoma)* which eats the barnacles.

Lower down the shore the variey of animals increases. The gobies here are the same species but are larger than the ones living in the higher pools and there are several other species of fish including a bright green wrasse with dark markings *(Thalassoma pavo)*; in close relations of this fish, studied elsewhere, it has been found that each group of individuals normally contains only a single male, but if he dies or is removed a large female changes sex and can then fertilize the eggs of the other females. You may also see a damselfish *(Abudefduf luridus)* which is bright blue when young, and strongly territorial, so that there is normally only one individual in a small pool. There is a tiny crab *(Pachygrapsus* sp.) which lives in holes, and pops out of sight when you walk near; a hermit crab *(Clibanarius aequabilis)* living in mollusc shells and also a crab *(Percnon planissimum)* which is very flat and seems to slither along the rocks. Much of the beauty of the pools results from the mauve, white, cream and pink of calcareous algae *(Lithophyllum* sp.) encrusting the rocks there, with the greenish anemones *(Anemonia sulcata)*, the occasional bright red of the sponges *(Clathrina coriacea)* and the tiny red or yellow of a coral *(Balanophyllia regia)*.

Among the molluscs that are common in the pools there is a worm shell *(Vermetus gigas)* which looks like a curled up calcareous worm, and an abalone *(Haliotis coccinea)* with a gently curved shell with a row of holes along one edge. There is also a black starfish *(Ophioderma longicauda)* and two common sea urchins, *Arbacia lixula* which is black and eats the calcareous algae, and *Paracentrotus lividus* which can change its colour from red to wine coloured to greenish black. If you are lucky you may come across a sea-hare *(Aplysia dactylomela)*, a large slug-like mollusc which is very variable in colour. On the rocks in some places there are "lawns" of a yellowish-white funnel-shaped alga *(Padina pavonica)*: this is a warm water species whose range extends only to the south coast of Britain. Among the algae you may find a shining green polychaete worm *(Eulalia viridis)*.

The deeper pools have a mass of yellowish-brown algae, *Cystoseira* spp. There are four species of this alga on the coast; one of these is only exposed at low tide, whilst the other three occur at different levels up the shore. These algae partially replace ecologically the various species of greenish-brown algae of the genus *Fucus* which characterize different levels on the rocky shores of northwest Europe, but which are much less dominant in the Canaries. One of the inhabitants of the deep pools that are exposed only at the lowest tides is the **Common Octopus** *(O.vulgaris* - known locally as **Pulpo**). You may see people hunting for these - often using one they have already caught as a lure to entice others out of their lairs.

SOME COMMON ANIMALS. See SELECTED MARINE FISH and SELECTED MARINE INVERTEBRATES, Section 3.

SOME COMMON ALGAE. *Cystoseira abies-marina* (mainly below the intertidal zone), *C.compressa* (further up), *C.humilis and C.discor* (in deep pools), *Padina pavonica* (funnel-shaped), *Lithophyllum* sp. (encrusting purple calcareous alga).

24. Ferry to *La Gomera*

Habitat description

Sea between Tenerife and La Gomera.

HOW TO GET THERE. *"FERRY GOMERA"* run a twice daily ferry service between *LOS CRISTIANOS* in the south of *TENERIFE* and the nearby island of *LA GOMERA*. There is a linking bus from *SANTA CRUZ* main bus station - see Appendix.

PRACTICAL DETAILS. Take binoculars and a warm jacket, however warm it is on shore. The ferry trip takes 80min. In winter the return trip is in the dark.

The excursion

The distance between the islands of Tenerife and the much smaller island of La Gomera is 39km (24mi). A trip on the ferry provides an opportunity to see animals that you are less likely to see from the shore, although there is obviously no guarantee that you will see anything at all! As far as invertebrates are concerned, the most striking are two species of floating jellyfish relatives - the **Portuguese Man-o'-war** *(Physalia physalis)* and the **By-the-wind-sailor** *(Velella spirans)* - which are quite common in these waters in spring; both species are very occasionally seen on the coasts of northwest Europe after prolonged southwest winds. The stings of these animals - like those of jellyfish - can be very painful indeed, so don't ever be tempted to touch one, even if washed up and apparently dead. They are preyed on by a nudibranch *(Glaucus atlanticus)*, a type of mollusc that floats on the sea surface.

Among vertebrate animals, four species of turtle occur in the sea around the islands, although there is no record of them breeding on the beaches here. If you get a glimpse of one of them it is most likely to be the **Loggerhead Turtle**. The **Common Dolphin** is often to be seen, and occasionally **Killer Whale** and **Great Sperm Whale**; seven other species of whales and dolphins are recorded from these waters. A large flock of feeding seabirds would probably indicate the presence of a shoal of fish or squid, and if you are lucky you might see swirls of activity in the water and catch a glimpse of one of the several species of tuna, or other predatory fish, driving smaller fish or squid to the surface. The small fish try to escape the large fish only to be caught by the seabirds - out of the frying pan into the fire.

If you keep a careful watch forwards during the trip you will have a fair chance of seeing **Flying Fish** *(Exocoetus volitans)*. These fish have enormously expanded pectoral fins, which enable them to glide as much as 100m (110yds) low over the water after being disturbed by predatory fish or by the ship. As they appear to be on the point of returning to the water, you may see them get an extra kick off with a powerful flick of the enlarged lower tail fin and continue on their flight.

For bird-watchers, this trip can be quite rewarding. As you leave Los Cristianos you will probably see the local dark-backed, yellow-legged **Herring Gulls**, which have a colony on cliffs nearby, and **Common Terns** are also often present. In summer the most conspicuous birds once you leave the shore will be the large grey-brown and white **Cory's Shearwaters**, magnificent long-winged gliders which frequently pass close across the bows or behind the stern of the ship but are not often to be seen feeding during this trip; they breed on the islands in summer and migrate south in winter. An occasional visitor in early autumn is the rather similar Great Shearwater *(Puffinus gravis)*, which breeds in the southern hemisphere; it can be distinguished from **Cory's Shearwater** by its "capped" appearance, dark bill, and much more prominent white band at the base of the tail. A much smaller species, the **Little Shearwater**, is a scarce local resident; although it breeds in winter a few can be seen at any time of the year. It is blackish above and white below like the more familiar Manx Shearwater *(Puffinus puffinus)*, but it is smaller, shorter-winged and with a decidedly chunky build; its flight is not graceful, and typically involves series of about 4 or 5 quick flaps followed by short glides.

Another local breeder that is likely to be unfamiliar to people from northwest Europe is **Bulwer's Petrel**; it is sooty-brown all over except for an indistinct pale bar across the inner part of the wing, and is the size of a small shearwater but with relatively longer wings and a long, pointed tail; the flight is buoyant and erratic. This species breeds in summer and moves south to winter in the tropics. Two other species that might be seen on this trip are the **Storm Petrel** and **Madeiran Storm-petrel**, which have both recently been found breeding in small numbers on islets off Tenerife. Both are blackish-brown with white rumps; the **Storm Petrel** is very small, with a white stripe on the underwing and weak, fluttering flight; the **Madeiran Storm-petrel** is larger, with all-dark underwing and relatively buoyant flight. Identification of these species is complicated, however, by the occurrence of several other kinds of storm petrels in Canarian waters. In winter a number of the resident species are absent but some visitors arrive from further north: Kittiwakes *(Rissa tridactyla)* and skuas (*Stercorarius* spp.) and Lesser Black-backed Gull *(Larus fuscus)* are regularly seen, especially on the approach to La Gomera.

Section 3. PLANTS AND ANIMALS

SELECTED PLANTS

The following list serves two purposes. It gives additional information on all the wild-growing conifers, flowering plants and ferns (referred to here simply as plants) that are mentioned in the text of SECTIONS 1 & 2, and it also includes other information that is not readily available elsewhere. Mosses are not included, and lichens are mentioned briefly at the end.

Nearly 200 plants (representing 70 families) are mentioned in the text; we give further details about all of these. Although this is only 14% of the wild-growing plants recorded from the island, nearly all the trees and the common and most conspicuous shrubs are mentioned, as well as other plants that are locally typical or especially interesting. Cultivated plants that do not grow wild or semi-wild are not included, even if they are mentioned in the text. Many plants should be recognizable from the thumbnail descriptions that we give, but this is not intended to be a detailed identification guide. We have, however, tried to include helpful field characters where appropriate, especially in the case of the euphorbias, the laurel forest trees and the common plants of the subalpine zone. We have kept the use of technical descriptive terms to a minimum; the three that should perhaps be defined are "palmate" for leaflets spreading out from one point like the palm of a hand as in Horse Chestnut, "trifoliate" for leaflets in threes as in Clover and "pinnate" for leaflets that arise from either side of a central stalk as in Ash. In many genera there are several species on the island but we mention this only when there are more than three. We give details in the APPENDIX of books that will help further with identification.

We could have left the list at that but, as discussed elsewhere in the book, one of the fascinations of Canary Island botany is the fact that a large proportion of plants are endemic (occur nowhere else). We decided to extract data from the lists published in FLORA OF MACARONESIA by A.Hansen and P.Sunding, and list here 123 out of the 137 families represented on Tenerife; we have not detailed 14 of the fern families. Under each family heading we mention the number of species that grow wild on Tenerife; there are 1331 of these (not including subspecies and hybrids). We also give, in brackets, the number of these species that are endemic to the Canary Islands; there are 286 of these.

There is much uncertainty as to which of the plants now growing apparently wild on the Canary Islands reached them naturally (ie. are native) and which were introduced by humans. The endemic species are clearly native, but the status of the non-endemic species is less obvious. It has been suggested that more than half of these were accidentally introduced, or escaped from cultivation, but some botanists now think that most of them reached the islands without the help of humans.

The families are listed in the order given in FLOWERING PLANTS OF THE WORLD, edited by V.H.Heywood, 1985; species are listed alphabetically by Latin name within each family. Latin names for plants, and also English or Spanish names if they are used in the text (see A NOTE ON NAMES AND TERMS, page 13) all appear in the index with the appropriate page reference to this section. Plant names follow FLORA OF MACARONESIA (mentioned above). We assume that most serious botanists on the island will have a copy of WILD FLOWERS OF THE CANARY ISLANDS by David and Zoë Bramwell for identifying endemic plants, and where a Latin name is different from the one used in their book, we include the latter in brackets. We have generally avoided mentioning subspecific variations. Where there is no established English name for a plant but it does have close relatives in Britain, or among familiar garden plants, we have sometimes mentioned the general English name. *FLORE DE L'AFRIQUE DU NORD* by René Maire, and FLOWERS OF THE MEDITERRANEAN by Oleg Polunin and Anthony Huxley, are both good sources of information on non-endemic plants.

Abbreviations and definitions

Status. Introduced; not endemic; Macaronesian endemic; Canary endemic; Tenerife endemic (see A NOTE ON NAMES AND TERMS on page 13).

Growth form. Tree; small tree; tall shrub - over 1.5m (5ft); small shrub - up to 1.5m; shrublet - up to 0.5m (20in); climber; fern; herbaceous perennial; woody-based herbaceous perennial; grass. The words scrambling, succulent and deciduous are also used.

Locality. CO - coast (used for plants typical of places influenced by sea spray); LO - lower zone; LF - laurel forest; PF - pine forest; HM - high mountain. This type of listing cannot be entirely precise; some plants may occur only in a limited part of the zone indicated, and some may extend outside the zone indicated. All the plants listed here can be found on at least one of the excursions.

Abundance. Widespread and abundant; widespread and frequent; widespread but scarce; locally abundant; locally frequent; local and scarce.

CONIFERS

Family **CUPRESSACEAE - cypresses, junipers** - 2 species on Tenerife
Juniperus cedrus, **Cedro Canario**.
>Macaronesian endemic; tree; HM local and scarce. Mainly on or near the cliff walls of Las Cañadas. Drooping branches; flat needle-like leaves; reddish-brown spherical cones in axils of leaves.

Juniperus phoenicea, **Sabina**, **Phoenician Juniper.**
>Not endemic; tree; LO local and scarce. More widespread in the past in the arid zone below the pine forest. Scale-like leaves close against twigs; dark red terminal cones. Also occurs elsewhere in Macaronesia and around the Mediterranean and on the west African coast.

Family **EPHEDRACEAE** - 3 species on Tenerife

Family **PINACEAE - pines** - 3 species on Tenerife (1 Canary endemic)
Pinus canariensis, **Pino Canario**, **Canary Pine**.
>Canary endemic (recorded as a fossil from Pliocene of southern Europe); tree; PF widespread and abundant. Needles up to 30cm long, in threes and densely crowded; cone 10-20cm long. Felling in the past has drastically reduced the area of natural Canary Pine forest; large areas are now replanted. There are two very large pines near the road just north of Vilaflor; one has a trunk diameter of 2.66m (8.72ft) and is 60m (197ft) high.

Pinus radiata, **Monterey Pine**.
>Introduced; tree; locally abundant. Needles 10-15cm long in threes or sometimes twos; cone 7-14cm long. Has been planted in the forests until recently, especially in the Orotava valley, but now the policy is to plant *Pinus canariensis*. A native of southern California.

FLOWERING PLANTS

Family **LAURACEAE - laurels, bays, avocados** - 4 species on Tenerife
>Four of the many evergreen trees of the laurel forest belong to this family. They all have alternate, broad-pointed and smooth-edged leaves, but can be hard to distinguish from trees belonging to other families that are also found in the forest.

Apollonias barbujana, **Barbusano.**
>Macaronesian endemic; evergreen small tree or tall shrub; LF locally frequent. Young leaves and twigs rosy; shiny leaves, edges untoothed and slightly curled, often with warts, but no glands; small greenish flowers; fruit an egg-shaped berry (up to 1.3cm). The leaves, when crushed, smell of mandarin. The only other member of this genus occurs in India, but fossils are known from the Mediterranean.

Laurus azorica, **Loro, Laurel.**

Not endemic; evergreen tree; LF widespread and abundant. Like a bay tree; dull untoothed leaves with a series of small inconspicuous glands or spots (not just a single pair of large ones) between the mid-rib and side veins; small greenish flowers; small black egg-shaped berry. In many places this is the dominant tree of the laurel forest. Found in other parts of Macaronesia and also SW Mediterranean and northern Africa.

Persea indica, **Viñátigo, an avocado.**

Macaronesian endemic; evergreen tree; LF locally frequent. Grows in damp places; large untoothed pale green leaves, somewhat limp and with no glands; most individuals have a few bright orange-red old leaves; small green flowers; fruit small (like a tiny avocado), not in a cup.

Ocotea foetens, Til, Tilo.

Macaronesian endemic; evergreen tree; LF local and scarce. Grows in damp places and not where the forest has been disturbed; large, shiny untoothed leaves with 2 large glands at the base of mid-rib and sometimes others along the mid-rib; small greenish flowers; fruit acorn-like with cup.

Family **ARISTOLOCHIACEAE** - 1 species on Tenerife

Family **RANUNCULACEAE - buttercups etc.** - 16 species on Tenerife
Ranunculus cortusifolius, **Morgallana, Morgallón.**

Macaronesian endemic; herbaceous perennial; LO,LF widespread and frequent. Extracts from this plant are used in herb medicine. Broad lobed leaves; large yellow 5-petalled buttercup flowers. There are 12 species of *Ranunculus* on Tenerife.

Family **PAPAVERACEAE - poppies etc.** - 10 species on Tenerife

Family **FUMARIACEAE - fumitories** - 8 species on Tenerife (1 Canary endemic)

Family **MYRICACEAE** - 1 species on Tenerife
Myrica faya, **Haya, Faya.**

Not endemic; tall evergreen shrub or tree; LF widespread and abundant, PF locally frequent. This evergreen is a close relative of Bog Myrtle; leaves rather similar to *Laurus* but smaller (although variable in size), without glands and usually with slightly serrated edges; they are duller and usually narrower than **Small-leaved Holly**; male plants have catkins, female plants have small black waxy fruit with a rough, almost faceted surface. Together with *Erica* spp. this plant forms the *fayal-brezal* (see Section 1. ECOLOGICAL ZONES - Laurel forest). It also occurs as understory in some parts of the pine forest. Also found elsewhere in Macaronesia and in southwest Iberia. The seeds have been used, in times of food shortage, to make a type of flour.

Family **FAGACEAE - beeches, oaks, chestnuts etc.** - 4 species on Tenerife
Castanea sativa, **Sweet Chestnut**.

Introduced; deciduous tree; in cultivated areas and often in dense stands close to laurel forest in north of the island. Rugged bark; long fairly broad toothed leaves; catkins; brown nuts in spiny jacket. In winter the bare trees appear lifeless and out of place alongside evergreen forests.

Family **CACTACEAE - cactuses** - 6 species on Tenerife
Opuntia spp., *Tunera*, **Prickly Pear**.

Introduced; succulent; LO locally abundant. 4 species are found on the island; stems are composed of flat pads with spines; yellow or red flowers with several petals fused into a cup. Cactuses are New World plants and are not part of the native flora of the Canary Islands. However *Opuntia* species are now well established; they thrive in disturbed areas and invade relatively undisturbed areas in several places, to the detriment of the native vegetation. Broken off "pads" take root to form new plants and the seeds are dispersed when birds, especially the Raven, feed on the fruit. In the past *Opuntia* was extensively cultivated as the food of the cochineal bug which was then collected for the production of the red cochineal dye.

Family **AIZOACEAE - mesembryanthemums etc.** - 5 species on Tenerife
Mesembryanthemum crystallinum, **Barrilla, Escarchosa**, **Ice Plant.**

Not endemic; succulent herb; CO locally abundant. Much branched and mat-forming; fleshy flat oval leaves covered with swollen crystalline hairs looking like water drops; white daisy-flowers; red succulent fruit. Seeds from this plant have been used for food in hard times in the past: they were toasted and then ground and used as *gofio*, which is still much eaten on the islands, but generally made from wheat or maize. Also occurs elsewhere in Macaronesia, in the Mediterranean and further south in the Old World.

Family **CARYOPHYLLACEAE - campions, pinks** - 47 species on Tenerife (13 Canary endemics)
Polycarpaea nivea, *Pata de Conejo.*

Not endemic; succulent shrublet; CO locally frequent. Prostrate; small succulent silvery leaves; small silvery flower heads. Also found elsewhere in Macaronesia (Cape Verde Islands) and in the southern Sahara.

Seven species of *Polycarpaea* can be found on Tenerife, 6 of which are Canary endemics.

Silene nocteolens, **Hierba Conejera, a campion**.

Canary endemic; small shrub; HM locally frequent. Lance-shaped leaves at base of flowering stem; groups of pinkish-white 5-petalled flowers on long stem. Thirteen species of *Silene* occur on Tenerife, 5 of which are Canary endemics.

Family **NYCTAGINACEAE - bougainvilleas etc**. - 1 species on Tenerife

Family **AMARANTHACEAE - cockscombs etc**. - 10 species on Tenerife (1 Canary endemic)

Family **PHYTOLACCACEAE - pokeweeds etc**. - 1 species on Tenerife

Family **CHENOPODIACEAE - beets, goosefoots** - 26 species on Tenerife (2 Canary endemics)

Atriplex glauca.

Not endemic; shrublet; CO locally frequent. Silvery alternate oblong toothed leaves, slightly fleshy; inconspicuous small 5-petalled flowers in spikes. Also occurs elsewhere in Macaronesia, the Mediterranean and northwest Africa.

Traganum moquinii, **Balancón**.

Not endemic; small shrub; CO local and scarce, in sandy places. Robust branched shrub; crowded fleshy leaves; inconspicuous hairy flowers. It also occurs elsewhere in Macaronesia (Cape Verde Islands) and in northern Africa.

Family **PORTULACACEAE** - 1 species on Tenerife

Family **BASELLACEAE** - 1 species on Tenerife

Family **POLYGONACEAE - docks, sorrels** - 15 species on Tenerife (1 Canary endemic)

Polygonum maritimum, **Sea Knotgrass**.

Not endemic; shrublet; CO locally frequent. Much branched low spreading shrublet with narrowly elliptical leaves; flowers pink or whitish, solitary or in small clusters; fruit a glossy nut. Widespread on coasts of both the Old and New World. There are 4 species of *Polygonum* on Tenerife.

Rumex lunaria, **Vinagrera, a dock**.

Canary endemic; small shrub; LO locally abundant. Compact bushy growth; glossy broad leaves; typical dock flower heads. Also occurs in disturbed places in forest zones. There are 8 species of *Rumex* on Tenerife, but *R. lunaria* s the only one that grows to shrub size.

Family **PLUMBAGINACEAE - sea-lavenders, thrifts** - 9 species on Tenerife (5 Canary endemics)

Limonium spp., *Siempreviva*, **Sea-lavender, Statice.**
Eight species (and several subspecies) on the island, nearly all Canary endemics; herbs or shrublets. Flower heads in branched clusters of one-sided spikes; flowers with mauve papery "everlasting" calyx.

Limonium pectinatum¹, Siempreviva de la Mar.
Macaronesian endemic; shrublet; CO widespread and abundant. Broad blunt-ended leaves usually in dense rosettes; flowers as above.

Family **THEACEAE - Tea, camellias** - 1 species on Tenerife

*Visnea mocanera, **Mocán.***
Macaronesian endemic species in a Macaronesian endemic genus; tree; LF local and scarce. Most likely to be found on cliffs; small elongate regularly toothed leathery leaves without glands; 5-petalled whitish flowers; edible lumpy fruit.

Family **GUTTIFERAE (HYPERICACEAE) - St. John's worts etc**. - 5 species on Tenerife (1 Canary endemic)

Hypericum spp., **St. John's worts**.
Five species of *Hypericum* occur on the island. They are shrubs with opposite leaves and open 5-petalled yellow flowers.

*Hypericum canariense, **Granadillo.***
Macaronesian endemic; tall shrub; LO,LF locally abundant. Long leaves; dense heads of flowers.

*Hypericum inodorum (H.grandifolium), **Malfurada.***
Macaronesian endemic; small shrub; LF,PF locally frequent. Reddish-brown stems; fairly broad leaves; few flowers in flower head.

*Hypericum reflexum, **Cruzadilla.***
Canary endemic; small shrub; LO,LF locally frequent. Small stalkless leaves crowded on stem; dense heads of flowers.

Family **TILIACEAE - limes etc** - 1 species on Tenerife

Family **MALVACEAE - hibiscus etc** - 15 species on Tenerife (2 Canary endemics)

*Lavatera acerifolia, **Malva de Risco.***
Canary endemic; tall shrub; LO locally frequent. A hibiscus relative; straggly shrub; fig-leaf-shaped leaves; pale mauve flowers with 5 free petals. There are 5 species of *Lavatera* on Tenerife, 2 of which are Canary endemics (1 on Tenerife only).

Family **ULMACEAE - elms etc** - 1 species on Tenerife.

Family **MORACEAE - figs, hemp, mulberries** - 2 species on Tenerife

Ficus carica, **Higuera, Fig**.

> Introduced; small deciduous tree; LO locally frequent. Cultivated but also found semi-wild in places that have previously been cultivated. Large lobed leaves; flower in fleshy pear-like structure which matures into the edible fig. Found throughout the Mediterranean.

Morus nigra, **Moral, Morena**, **Mulberry.**

> Introduced; tree; LO local and scarce. Broad 3-lobed toothed leaves; flowers in short catkins; clusters of dark red fruit. Occasionally found semi-wild; also used as shade tree.

Family **URTICACEAE - stinging nettles etc.** - 10 species on Tenerife (4 Canary endemics)

Forsskaolea angustifolia, **Ratonera.**

> Canary endemic; shrublet; LO locally frequent. Slender stems with very small spiny thistle-like leaves, woolly below; small pinkish flowers. Common roadside plant. Occurs in Africa.

Gesnouinia arborea, **Ortigón de Monte.**

> Canary endemic species in a Canary endemic genus; tall shrub; LF local and scarce. Nettle relative; simple fairly long leaves covered in fine hairs; terminal clusters of small flowers in spikes.

Urtica morifolia, **Ortigón, a stinging nettle.**

> Macaronesian endemic; woody perennial; LF locally frequent. Toothed leaves with stinging hairs; small clusters of flowers in leaf axils.

Family **VIOLACEAE - violets, pansies** - 7 species on Tenerife (2 Canary endemics)

Viola cheiranthifolia, **Violeta del Teide, Teide Violet.**

> Canary endemic (Tenerife only); herbaceous perennial; HM local and scarce. Fairly long smooth-edged leaves; solitary lilac-coloured pansy-like flowers. Grows on Mt.Teide up to about 3500m (11500ft) - one of the highest naturally occurring plants on the island. A total of seven species of *Viola* can be found on Tenerife (2 on Tenerife only).

Family **CISTACEAE - rockroses** - 9 species on Tenerife (4 Canary endemics)

Cistus monspeliensis, **Jara, Juagarzo.**

> Not endemic; small shrub; upper LO,PF locally frequent. Sticky narrow leaves; small white 5-petalled flowers. Most frequent in the lower parts of the pine forest, especially in areas where there have been forest fires.

Also occurs elsewhere in Macaronesia, in the Mediterranean and northwest Africa.

Cistus osbaeckiaefolius, **Jara de las Cañadas.**

Canary endemic (Tenerife only); small shrub; HM local and scarce. Lance-shaped leaves, smaller and hairier than in the following species; pink 5-petalled flowers.

Cistus symphytifolius, **Amagante, Jarón.**

Canary endemic; small shrub; PF locally abundant. Lance-shaped hairy leaves; large, somewhat wrinkled, pink 5-petalled flowers. A dominant plant of the understory in many areas of pine forest, especially at high levels; it is favoured by the passage of forest fires.

Family **TAMARICACEAE - tamarisks** - 2 species on Tenerife

Tamarix canariensis, **Tarajal, Tamarisk.**

Not endemic; small shrub; CO locally frequent. Reddish brown bark; scale-like leaves; minute pink flowers in slender spikes; seeds with tufts of hairs. Also occurs elsewhere in Macaronesia (Cape Verde Islands), and in southern Europe, northern Africa and Asia.

Family **FRANKENIACEAE - frankenias** - 3 species on Tenerife

Frankenia laevis, **Sea-heath.**

Not endemic; shrublet; CO widespread and frequent. Mat-forming plant; opposite linear leaves; terminal clusters of pink or whitish 5-petalled flowers. Also occurs elswhere in Macaronesia, western Europe, the Mediterranean and northwest Africa.

Family **CUCURBITACEAE - gourds, pumpkins** - 3 species on Tenerife (1 Canary endemic)

Family **SALICACEAE - willows, aspens, poplars** - 3 species on Tenerife

Salix canariensis, **Sauce, Sao, Canary Willow.**

Macaronesian endemic; tall shrub or tree; LO,LF,HM locally frequent. In damp places; elongate alternate leaves, often with galls; catkins.

Family **CRUCIFERAE (BRASSICACEAE) - mustards, cabbages etc**. - 58 species on Tenerife (14 Canary endemics)

Crambe arborea, **Col de Risco.**

Canary endemic (Tenerife only); small shrub; LO local and scarce. On cliffs; slender ridged stem; alternate fairly broad very jagged leaves; 4-petalled white flowers. There are 5 species of *Crambe* on Tenerife, all of which are Canary endemics (two are found only on Tenerife).

Descurainia bourgeauana, **Hierba Pajonera.**

Canary endemic (Tenerife only); small shrub; HM widespread and abundant. Sheep-size clumps; densely leafy; pinnate leaves; long spikes of 4-petalled yellow flowers. One of the commonest plants in Las Cañadas. Five species of *Descurainia* are found on Tenerife; they are all Canary endemics and four occur only on Tenerife.

Erysimum scoparium (Cheiranthus scoparium), **Alhelí del Teide.**
Canary endemic (Tenerife only); small shrub; HM widespread and abundant. Linear leaves; 4-petalled mauvish-pink mustard-like flowers. One of the commonest plants in Las Cañadas.

Parolinia intermedia.
Canary endemic (Tenerife only); small shrub; LO locally frequent. Erect compact shrub; narrow leaves; 4-petalled pinkish flowers.

Family **RESEDACEAE - mignonettes etc**. - 3 species on Tenerife (1 Canary endemic)
Reseda scoparia.
Canary endemic; small shrub; LO locally abundant. Linear fleshy leaves; small white flowers on spike.

Family **ERICACEAE - heaths, rhododendrons** - 4 species on Tenerife (1 Canary endemic)
Arbutus canariensis, **Madroño, a strawberry tree.**
Canary endemic; evergreen small tree; LF,PF locally frequent. Bark russet-coloured and peeling; elongate regularly toothed leaves, quite unlike more familiar *Erica*, more like a laurel; flowers very large but heather-like; fruit 2-3cm yellow/orange edible berry.

Erica arborea, **Brezo, a tree heath.**
Not endemic; small evergreen tree or shrub; LF widespread and abundant, PF locally frequent. A giant heath or heather; hard, finely grooved bark; linear leaves more erect than *E.scoparia*; white flowers. Also found elsewhere in Macaronesia, Mediterranean and northern Africa.

Erica scoparia subsp. *platycodon*, **Tejo, a tree heath.**
Macaronesian endemic subspecies; tall evergreen shrub; LF locally abundant, especially on high ridges. A giant heath or heather; bark tends to peel off in strips; linear leaves more spreading and geometrically arranged than in *E.arborea*; reddish-pink flowers.

Family **SAPOTACEAE** - 1 species on Tenerife
Sideroxylon marmulano, **Marmolán.**
Macaronesian endemic; small tree; LO,LF local and scarce. Long leathery untoothed glossy leaves; small white flowers crowded at base of leaves; fruit an egg-shaped berry. Closest relatives around the Indian Ocean.

Family **PRIMULACEAE - primroses etc**. - 5 species on Tenerife

Family **MYRSINACEAE** - 2 species on Tenerife (1 Canary endemic)
Pleiomeris canariensis, **Delfino.**

Canary endemic; small evergreen tree; LF local and scarce. Alternate large (over 20cm long) broad untoothed leaves (broader than *Heberdenia*); flowers with 5 petals united at base to form small tube; round pink fruit 5-7mm across.

Heberdenia excelsa (H.bahamensis), **Sacatero, Aderno.**

Macaronesian endemic; evergreen tree; LF local and scarce or frequent. Whitish trunk; alternate broad untoothed leaves (not so broad as *Pleiomeris*); 5-petalled strongly scented white flowers; hard round berry 5mm across. This species has suffered from exploitation, although a few groups of mature trees can still be found. However, it is now growing back in some areas and forming part of the undergrowth.

Family **PITTOSPORACEAE** - 1 species on Tenerife

Family **ROSACEAE - roses etc**. - 15 species on Tenerife (2 Canary endemics)
Bencomia caudata.

Macaronesian endemic species in a Macaronesian endemic genus; small shrub; LF local and scarce. Toothed pinnate leaves; 5-petalled flowers in spike. Closest relatives in South Africa.

Marcetella moquiniana, **Palo de Sangre.**

Macaronesian endemic species in a Macaronesian endemic genus; small shrub; LO local and scarce. Upper part of stem is hairy and reddish; leaves in rosettes at end of stem; each leaf has a series of paired toothed leaflets; 5-petalled flowers in drooping spikes (separate male and female spikes).

Prunus lusitanica subsp. *hixa,* **Hija.**

Macaronesian endemic subspecies; evergreen tree; LF locally abundant (in Anaga, but not in Teno). Pinkish leaf stalks; fairly elongate glossy regularly toothed leaves; 5-petalled white flowers in long drooping spikes; blackish fruit. The subspecies also occurs in Madeira; the species occurs in southwest Europe.

Rubus spp., **Zarza,** Bramble.

Some Macaronesian endemics; scrambling shrubs; LO,LF locally abundant. Prickly stems; leaves with three leaflets; 5-petalled flower; fruit a blackberry. A difficult genus to sort out but probably two species are present on the island. Often form dense thickets, especially in disturbed areas.

Sorbus aria, **Peralillo de Cumbre, Whitebeam.**

Not endemic; small tree; HM local and scarce. Broad alternate leaves with regularly toothed edges and silvery-white underneath; white 5-petalled flowers in fairly dense flat heads; red fruit. Also occurs in Europe, northern Africa and further east. Occurs in a few places on the rim of Las Cañadas.

Family **CRASSULACEAE - stonecrops, houseleeks** - 41 species on Tenerife (35 Canary endemics)

Aeonium spp., *Bejeques.*

14 species on the island, all of them endemic to the archipelago and several to Tenerife, and also many subspecies and hybrids; succulent herbaceous perennials and shrublets; all zones locally frequent. Broad fleshy leaves in rosettes; leaves generally have fine hairs along the edges; many-parted star-like flowers in spike which rises from centre of rosette; species often hard to distinguish.

Aeonium canariense.

Canary endemic (Tenerife only); succulent woody-based herbaceous perennial; LO,LF locally frequent. On cliffs and rocks; broad fleshy leaves; yellow flowers.

Aeonium lindleyi, **Higureta.**

Canary endemic; succulent small shrub; LO locally frequent. Rocky places; thin woody branches; forms thickets; very thick blunt leaves; yellow flowers.

Aeonium spathulatum, **Bejeque.**

Canary endemic; succulent shrublet; LF,PF,HM locally frequent. On cliffs and rocks; woody branching shrublet; small spoon-shaped sticky leaves; yellow flowers.

Aeonium tabulaeforme, **Pastel de Risco.**

Canary endemic (Tenerife only); succulent woody-based herbaceous perennial; LO locally abundant. On cliffs and rocks; large plate-like rosettes of leaves (up to 30cm in diameter) growing flat against the rock; yellow flowers.

Aeonium urbicum, **Bejeque de los Tejados.**

Canary endemic; succulent small shrub; LO locally frequent. Broad fleshy whitish-green leaves in rosettes at the top of an unbranched stem; large flowering head with pink to greenish flowers. (This is the *Aeonium* frequently seen on roof-tops.)

Greenovia spp., *Bejeque, Pastel de Risco.*

4 species on Tenerife; succulent shrublets; all zones locally frequent. Rosettes of broad fleshy leaves; many-parted star-like flowers. Very

similar to *Aeonium*, but with smooth leaves which generally have a whitish bloom.

Monanthes spp.

Members of a Canary endemic genus (one species on the Salvages). 13 species on the island and several hybrids. Tiny succulent shrublets; all zones; hard to distinguish; grow in cracks in rocks; very small leaves in rosettes; many-parted star-shaped flowers.

Monanthes adenoscepes, **Arroz.**

Canary endemic (Tenerife only); succulent shrublet; HM locally frequent. Unbranched rosettes of densely crowded small succulent leaves; purple flowers.

Monanthes niphophila.

Canary endemic; succulent shrublet; HM local and scarce. Solitary rosettes of pointed leaves; greenish flowers.

Monanthes silensis.

Canary endemic; succulent shrublet: HM locally frequent. Dense clumps; branched rosettes of small pale green succulent leaves.

Family **HYDRANGEACEAE - hydrangeas** - 1 species on Tenerife

Family **LEGUMINOSAE (FABACEAE) - peas, gorse, broom** - 111 species on Tenerife (25 Canary endemics)

Adenocarpus foliolosus, **Codeso.**

Canary endemic; small shrub; LF,PF locally frequent. Spreading shrub with tiny densely packed trifoliate leaves along the stems; terminal spikes of yellow pea-flowers; similar to *Teline*.

Adenocarpus viscosus, **Codeso de la Cumbre, Codeso del Pico, Codeso del Teide**.

Canary endemic; small shrub; HM widespread and abundant. Shrub with sticky flowers and seed pods, otherwise hard to distinguish from *A.foliolosus* (with which it apparently sometimes hybridizes) except by locality. One of the commonest shrubs of Las Cañadas.

Chamaecytisus proliferus, **Escobón.**

Canary endemic; tall shrub; LO,PF,HM locally abundant. Straggly shrub; trifoliate leaves; white pea-flowers in small clusters at base of leaves; seeds in pods. The dominant understory plant in many areas of pine forest.

Lotus campylocladus, **Corazoncillo**, **a birdsfoot trefoil.**

Canary endemic; woody-based perennial; PF,HM locally frequent. Procumbent; grows larger in pine forest; leaves with 5 narrow leaflets, densely hairy; small clusters of yellow pea-flowers. 15 species of *Lotus* can be found on Tenerife (9 are Canary endemics).

Lotus sessilifolius, **a birdsfoot trefoil.**

Canary endemic; woody-based perennial; LO locally frequent. Leaves with 5 leaflets covered with silky hairs; small clusters of yellow pea-flowers either at base of leaves or in spike.

Ononis serrata, **a rest-harrow.**

Not endemic; annual herb; LO locally frequent. Sticky low growing plant with trifoliate leaves, leaflets having toothed margins; dense terminal clusters of white or pale pink flowers. Also occurs in northern Africa and southwest Asia. Of the 8 species of *Ononis* on Tenerife only one is a Canary endemic.

Retama raetam (R.monosperma), **Retama, Retama Blanca.**

Not endemic; shrub; LO locally frequent. Erect and much branched shrub; small trefoil leaves with 3 leaflets; white pea-flowers; pods with only one or two seeds (cf. *Spartocytisus supranubius*).

Spartocytisus supranubius, **Retama del Teide, Retama Blanca, Retama del Pico** (**Teide Broom** in some tourist literature).

Canary endemic; tall shrub; HM widespread and abundant. Tall dense shrub; straight twigs and small trifoliate leaves; fragrant white or sometimes pinkish pea-flowers; pods with 4-6 seeds (cf. *Retama raetam*). One of the dominant plants of Las Cañadas.

Teline canariensis, **Gildana.**

Canary endemic; tall shrub; LF locally frequent. Compound dark green trifoliate leaves, often silvery below; flower spikes terminal; yellow pea-flower similar to *Adenocarpus* but with silky seed pods. Of the 6 species of *Teline* on Tenerife, 4 are Canary endemics.

Family **CAESALPINACEAE** - 4 species on Tenerife

Family **MIMOSACEAE** - 3 species on Tenerife

Family **THELIGONACEAE** - 1 species on Tenerife

Family **HALORAGACEAE** - 1 species on Tenerife

Family **LYTHRACEAE** - 2 species on Tenerife

Family **THYMELAEACEAE - daphnes etc** - 2 species on Tenerife

Daphne gnidium, **Trobisca.**

Not endemic; small evergreen shrub; PF locally abundant. Slender erect branches; small greyish-white leaves all the way up the branches; terminal clusters of white 4-petalled flowers; red berries. Also occurs in the Mediterranean and northwest Africa.

Family **MYRTACEAE - eucalyptus, myrtles** - 3 species on Tenerife
Eucalyptus spp., *Eucalipto.*

Introduced trees; locally abundant; common along roadsides and in small plantations near the top of the lower zone and extending into the forests. Can grow very large; flaky grey bark; strong eucalyptus scent; slender drooping leaves. Various species of these evergreen Australian trees have been grown extensively for timber and shade; two species now occur in semi-wild state.

Family **PUNICACEAE - pomegranates** - 1 species on Tenerife

Family **ONAGRACEAE - evening primroses, fuchsias** - 9 species on Tenerife

Family **SANTALACEAE - sandalwoods** - 4 species on Tenerife (2 Canary endemics)

Family **RAFFLESIACEAE** - 1 species on Tenerife

Family **CELASTRACEAE - spindle trees etc.** - 1 species on Tenerife (Canary endemic)
Maytenus canariensis, *Peralillo, Peralito.*

Canary endemic; tree; LO,LF locally frequent. Steep places; densely branched; small broad glossy slightly toothed leaves; pale green 5-lobed flowers; fruit is a pale green to light brown dry 3-celled capsule (cf. *Rhamnus crenulata*). Lack of prickles on leaves help distinguish it from *Rubia fruticosa.*

Family **AQUIFOLIACEAE - hollies etc.** - 2 species on Tenerife
Ilex canariensis, *Acebiño*, **Small-leaved Holly.**

Macaronesian endemic; evergreen tree; LF widespread and abundant. Glossy green alternate broad leaves without glands, slightly rolled down at the edges and lacking prickles (the occasional leaf has a few); white 4-petalled flowers; rather few red fruit 1cm across. We have given English names to the two hollies in order to distinguish them easily in the text, but this is the only case in this book where we have 'invented' names.

Ilex perado subsp. *platyphylla*, *Naranjero Salvaje, Acebo*, **Large-leaved Holly**.

Canary endemic subspecies; evergreen tree; LF locally frequent (in Anaga, but not in Teno). In damp places; large prickly glossy leaves; pinkish-white 4-petalled flowers; blackish fruit 6-9mm across. See note in previous entry.

Family **EUPHORBIACEAE - spurges etc.** - 29 species on Tenerife
(4 Canary endemics)

Euphorbia spp.

Stout-stemmed shrubs and shrublets with leaves (when they have them)
in terminal rosettes; sap is a poisonous milky latex. They are the
dominant plants in much of the lower zone. There are 27 species (and
some hybrids and subspecies) recorded from Tenerife.

Euphorbia aphylla, ***Tabaiba Parida, Tabaiba Salvaje, Tolda.***

Canary endemic; succulent shrublet; LO locally frequent. Cylindrical
stems which are erect and entirely leafless; smaller than other euphorbias; denser clumps and more slender stems than *Ceropegia*, with which
it could be confused.

Euphorbia atropurpurea, ***Tabaiba Majorera.***

Canary endemic; small shrub; LO locally abundant. Brown branched
stems with leaves clustered at tips; red/purple multiple flowers.

Euphorbia balsamifera, ***Tabaiba Dulce.***

Not endemic; tall shrub; CO,LO locally abundant. Forms large domes;
grey gnarled branching stems; terminal rosettes of leaves; single flowers; brown seed capsule. A good competitor in damp salty air. Also
occurs in Africa.

Euphorbia bourgeauana.

Canary endemic (Tenerife only); small shrub; LO locally frequent.
Broad short leaves in terminal rosettes; multiple greenish-yellow
flowers.

Euphorbia canariensis, ***Cardón.***

Canary endemic; tall succulent shrub; LO locally abundant. Grows in
large clumps (up to about 3m (10ft) high); pale green square candelabra-
like stems, becoming silvery with age, with a row of paired spines at
each corner; no leaves; terminal clusters of red flowers. This is not a
cactus, although it is succulent and spiny.

Euphorbia obtusifolia (E.regis-jubae), ***Tabaiba, Tabaiba Amarga.***

Not endemic; tall shrub; LO widespread and abundant. Erect, many-
branched shrub growing up to about 2m high, with brownish stems; terminal rosettes of slender leaves, multiple greenish flowers; reddish seed
capsules. Also occurs elsewhere in Macaronesia (Salvage Islands).

Euphorbia paralias, **Sea Spurge.**

Not endemic; succulent shrublet; CO local and scarce. Small, unbranched, with fleshy leaves overlapping up the stem. Also occurs elsewhere
in Macaronesia, in western Europe, the Mediterranean and northwest
Africa.

Ricinus communis, **Tártago, Castor Oil Plant.**
Introduced; tall poisonous shrub; LO widespread and frequent. Large broad palmate leaves; clusters of red flowers; burr-like fruit. Conspicuous along roadsides, especially in the north of the island. Native of the tropics.

Family **RHAMNACEAE - buckthorns etc**. - 3 species on Tenerife (2 Canary endemics)

Rhamnus crenulata, **Espinero, Espino Negro, Leña Negra.**
Canary endemic; small shrub; LO locally frequent. Small broad leaves with bluntly toothed edges and no glands; lack of prickles on leaves help to distinguish it from *Rubia;* fruit is single seeded (cf. *Maytenus*).

Rhamnus glandulosa, **Sanguino.**
Macaronesian endemic; small tree; LF locally frequent. Occurs in damp places; small broad shiny regularly toothed leaves with two or more conspicuous swollen glands along midrib; usually with pink leaf stalks; small 5-parted flowers; red berries, turning black.

Rhamnus integrifolia, **Moralito.**
Tenerife endemic; small shrub; LO,HM locally scarce. On cliffs; smooth-edged broad leaves with no glands; small 5-parted flowers; reddish-black 4-lobed fruit. Probably more widely distributed in the past.

Family **VITACEAE - grapevines** - 1 species on Tenerife

Family **SAPINDACEAE - litchis etc**. - 1 species on Tenerife

Family **ACERACEAE - maples** - 1 species on Tenerife

Family **ANACARDIACEAE - cashews etc**. - 3 species on Tenerife

Pistacia atlantica, **Almácigo.**
Not endemic; tree; LO,LF local and scarce. Pinnate untoothed leaves; small brownish flowers without petals, oval red berries. Also occurs in the Mediterranean and northwest Africa.

Family **SIMAROUBACEAE - quassias etc**. - 1 species on Tenerife

Family **CNEORACEAE - spurge olives etc**. - 1 species on Tenerife (Canary endemic)

Neochamaelea pulverulenta (Cneorum pulverulentum), **Leña Blanca, Leña Santa, Leña Buena, Orijama.**
Canary endemic; small shrub; LO locally abundant. A small densely hairy shrub; linear leaves; 3 or 4-parted yellow flowers on the leaf stalks; fruits with globular lobes. A common plant in euphorbia communities.

Family **RUTACEAE - citrus fruits** - 2 species on Tenerife (1 Canary endemic)

Family **ZYGOPHYLLACEAE - lignum vitaes etc**. - 3 species on Tenerife
*Zygophyllum fontanesii, **Uvas de Mar, Uvilla de Mar.***
 Not endemic; succulent shrublet; CO locally abundant. Adapted to dry coastal conditions; grape-like yellowish-green leaves. Also occurs elsewhere in Macaronesia and in northern Africa.

Family **JUGLANDACEAE - walnuts, pecans** - 1 species on Tenerife

Family **LINACEAE - flax, linseeds** - 5 species on Tenerife

Family **GERANIACEAE - geraniums etc**. - 16 species on Tenerife (2 Canary endemics)
*Geranium canariense, **Pata de Gallo**, **a cranesbill.***
 Canary endemic; woody-based herbaceous perennial; LF locally frequent. Robust herb; very dissected leaves in rosettes on stems; branched flower stems with 5-petalled pink flowers (2-3cm across). There are 5 species of *Geranium* on Tenerife, but only *G.canariensis* is a Canary endemic.

Family **OXALIDACEAE - wood sorrels** - 5 species on Tenerife

Family **BALSAMINACEAE - balsams** - 1 species on Tenerife

Family **TROPAEOLACEAE - nasturtiums** - 1 species on Tenerife

Family **ARALIACEAE - ivies etc**. - 1 species on Tenerife
Hedera helix subsp. canariensis (H.canariensis), **Hiedra, Yedra, Ivy.**
 Macaronesian endemic subspecies; climber; LF locally frequent. Looks much like mainland Europe subspecies which is also recorded from the island.

Family **UMBELLIFERAE (APIACEAE) - carrots etc**. - 38 species on Tenerife (9 Canary endemics)
*Astydamia latifolia, **Lechuga de Mar.***
 Not endemic; herbaceous perennial; CO, locally frequent. Fleshy jagged yellowish leaves; flat head of small yellow 5-petalled flowers. Also occurs elsewhere in Macaronesia (Salvage Islands) and in northern Africa.
*Bupleurum salicifolium, **a hare's-ear.***
 Macaronesian endemic; small shrub; LO locally frequent. Long narrow leaves; flat head of small yellow 5-petalled flowers.

Crithmum maritimum, **Perejil de Mar, Rock Samphire.**
Not endemic; herbaceous perennial; CO locally frequent. Pinnate leaves with narrow fleshy lobes; flat heads of small 5-petalled yellow flowers. Also occurs elsewhere in Macaronesia, in Europe and in northwest Africa.

Ferula linkii, **Cañaheja, Cañalé.**
Canary endemic; herbaceous perennial; LO,HM locally frequent. A giant fennel-like plant, growing up to about 3m (10ft) high; feathery leaves; dramatic flat heads of 5-petalled yellow flowers.

Foeniculum vulgare, **Hinojo, Fennel.**
Not endemic; herbaceous perennial; LO locally frequent. Strong smelling; feathery leaves; flat heads of 5-petalled yellow flowers. Also found elsewhere in Macaronesia and the Mediterranean.

Pimpinella cumbrae, **Perejil Blanco.**
Canary endemic; woody-based herbaceous perennial; HM locally frequent. Ridged branches; pinnate leaves with toothed edges; flat heads of small 5-petalled white flowers.

Tinguarra cervariaefolia.
Canary endemic species in a Canary endemic genus; herbaceous perennial; LO local and scarce. On cliffs; jagged pinnate leaves; flat heads of small white 5-petalled flowers. There are only two species in this genus, and both occur on Tenerife.

Todaroa aurea.
Canary endemic; herbaceous perennial; LO local and scarce. Pinnate leaves; flat heads of small yellow 5-petalled flowers.

Family **APOCYNACEAE - periwinkles etc**. - 2 species on Tenerife

Family **GENTIANACEAE - gentians** - 2 species on Tenerife (1 Canary endemic)

Ixanthus viscosus, **Reina del Monte.**
Canary endemic; herbaceous perennial; LF locally frequent. A tall (0.75m) gentian; almost stalkless opposite lance-shaped leaves with a few parallel veins; branched flowering heads with bell-shaped yellow 5-lobed flowers.

Family **ASCLEPIADACEAE - milkweeds etc**. - 6 species on Tenerife (3 Canary endemics)

Ceropegia dichotoma, **Cardoncillo.**
Canary endemic (Tenerife only); small succulent shrub; LO locally abundant on north coast. Fat upright greyish cylindrical stems; no leaves; strange yellow flowers, with 5 partly fused petals (cf. description of *Euphorbia aphylla*).

Ceropegia fusca, **Cardoncillo, Mataperro.**

Canary endemic; small succulent shrub; LO locally frequent in the south. Similar to *C.dichotoma* but with reddish brown flowers.

Periploca laevigata, **Cornical.**

Not endemic; scrambling shrub; LO locally frequent. Opposite fairly broad leaves; flowers green above, white and brownish inside, with 5 partially fused petals; strikingly long twin-horned seed pods which produce a mass of white fibres when the seeds ripen. Also occurs elsewhere in Macaronesia (Salvage Islands), in the Mediterranean and in northwest Africa.

Family **OLEACEAE - olives, ashes, lilacs** - 5 species on Tenerife (1 Canary endemic)

Jasminum odoratissimum, **Jazmín.**

Macaronesian endemic; tall shrub; LO locally frequent. Spreading shrub; alternate glossy pinnate leaves; small trumpet-shaped yellow flowers, fading to white; small oblong brown-black fruit, the size of a coffee bean. This species is typical of the transition from the upper part of the lower zone to the humid laurel forest zone.

Picconia excelsa, **Palo Blanco.**

Macaronesian endemic species in a Macaronesian endemic genus; tree or tall shrub; LF locally frequent. Whitish bark and somewhat warty branches; rather leathery broadly lance-shaped opposite leaves with untoothed edges, often slightly curled and without glands; 4-lobed white flowers; fruit black and olive-shaped. Closest relatives are in Eastern Australia, 19000km (11800mi) away, but fossils have been found in southern Europe.

Olea europaea subsp. *cerasiformis*, **Acebuche, Olive.**

Canary endemic subspecies; tall shrub; LO locally frequent. Grows on cliffs; grey bark; lance-shaped narrow untoothed opposite leaves which are glossy green above and silky white below; clusters of small whitish flowers with 5 partially fused petals; fruit - olive. Species also occurs elsewhere in Macaronesia, in the Mediterranean and in northwest Africa.

Family **SOLANACEAE - potatoes, nightshades** - 31 species on Tenerife (2 Canary endemics)

Datura stramonium, **Thorn-apple.**

Not endemic; annual herb (up to 1m); LO local and frequent. Luxuriant stout poisonous plant with large broad leaves; large white trumpet-shaped 5-lobed flowers; fruit egg-shaped, spiny and green, like a horse-chestnut. Now almost worldwide distribution.

Nicotiana glauca, **Bobo, Mimo, Tabaco Moro, Tree Tobacco**.

Introduced; tall shrub; LO widespread and frequent. Sparse straggly growth; large fairly broad leaves; flowers all the year round; yellow tubular flowers with 5 shallow lobes. A South American plant which thrives here and is especially frequent in disturbed ground such as along roadsides. There are 5 species of *Nicotiana* on Tenerife.

Withania aristata, **Orobal.**

Not endemic; tall shrub; LO locally frequent. Large broad leaves; bell-shaped 5-lipped yellowish-green flowers; small blackish berries. Also occurs in northern Africa.

Family **CONVOLVULACEAE - convolvulus, bindweeds** - 19 species on Tenerife (6 Canary endemics)

Convolvulus canariensis, **Correguelón.**

Canary endemic; scrambling shrub; LF locally frequent. Alternate oblong densely hairy leaves; pale blue or mauve trumpet-shaped flowers. There are 12 species of *Convolvulus* on Tenerife (6 Canary endemics).

Convolvulus floridus, **Guaydil.**

Canary endemic; tall shrub; LO locally abundant. Alternate oblong leaves; dramatic sprays of white trumpet-shaped flowers. Common garden shrub.

Family **HYDROPHYLLACEA** - 1 species on Tenerife

Family **BORAGINACEAE - borages, forgetmenots, buglosses** - 24 species on Tenerife (12 Canary endemics)

Echium spp., **Taginaste**, **bugloss.**

Macaronesian endemic genus, mainly Canary Islands. 12 species in Tenerife; 11 of these are Canary endemics (5 on Tenerife only). Shrubs or herbs; all zones, locally frequent. Alternate simple narrow leaves; spikes of tube-like flowers with 5 lobes.

Echium aculeátum.

Canary endemic; tall shrub; LO locally frequent (only in the west). Branched shrub; linear spiny leaves; many compact groups of white flowers.

Echium áuberianum, **Tajinaste Picante.**

Canary endemic (Tenerife only); woody-based herbaceous perennial; HM local and scarce. Basal rosette of linear leaves; slender spike of blue flowers.

Echium giganteum.

Canary endemic (Tenerife only); tall shrub; LO,LF locally frequent.

Stem branched; long lance-shaped hairy leaves; cone-shaped clusters of white flowers.

Echium wildpretii, **Tajinaste Rojo.**

Canary endemic; short-lived perennial; HM local and scarce. Unbranched; dense rosette of linear leaves; single spectacular flowering stem up to several metres tall, growing from centre of leaf rosette; tightly packed red flowers.

Myosotis latifolia, **a forgetmenot.**

Not endemic; woody based herbaceous perennial; LF locally frequent. Tall forgetmenot; flowers usually blue. Also found elsewhere in Macaronesia and in northwest Africa.

Family **VERBENACEAE - teaks, verbenas** - 6 species on Tenerife

Family **LABIATAE (LAMIACEAE) - sages, dead-nettles, thymes** - 59 species on Tenerife (25 Canary endemics)

Members of this family are often aromatic; they have square stems, opposite leaves and tubular two-lipped flowers.

Bystropogon spp., **Poleo.**

This is a genus which occurs only in the Canaries, Madeira and western south America! There are 4 species on the island, which are quite hard to distinguish. The leaves are used for making herb tea.

Bystropogon canariensis, **Poleo del Monte.**

Canary endemic; tall shrub; LF locally frequent. Tall aromatic shrub, but not smelling of mint; broad toothed leaves which are green above; clusters of small round pink and white flowers.

Bystropogon origanifolius, **Poleo.**

Canary endemic; small shrub; LF,PF,HM locally frequent. Aromatic shrub smelling of mint; broad leaves silvery above and almost toothless; clusters of small 2-lipped white flowers.

Cedronella canariensis, **Algaritofe.**

Macaronesian endemic; woody-based herbaceous perennial; LF locally frequent. Scented, trifoliate leaves with lance-shaped leaflets; dense spike of pink flowers.

Lavandula spp., **lavenders.**

5 species on Tenerife and some hybrids; small shrubs; LO, locally abundant. Deeply dissected leaves (cf. *Nepeta.*); blue flowers in short spikes.

Micromeria spp., **Tomillo.**

7 species and some hybrids on the island; hard to distinguish; 4 species endemic to Tenerife, the others endemic to Canaries or Macaronesia; shrublets; all zones. Small leaves; tiny thyme-like flowers.

Nepeta teydea, **Hierba del Teide, Tonática, a catmint.**
Canary endemic; herbaceous perennial; HM locally frequent. Long simple toothed leaves help distinguish it from *Lavandula*; purple flowers in spike.

Salvia canariensis, **Salvia, a sage.**
Canary endemic; small shrub; LO locally frequent. Large almost triangular pointed leaves with whitish bloom; pink flowers.

Salvia broussonetii, **a sage.**
Canary endemic; small shrub; LO, locally frequent. Leaves large, very broad, not pointed, coarsely toothed; flowering head usually branched; flowers tubular, with upper lip hooded, lower 3-lobed, white to pinkish.

Sideritis spp., **Chahorra, Salvia Blanca.**
12 endemic species on Tenerife, some with subspecies; small shrubs and shrublets; all zones widespread and frequent. Sage-like; stems and branches hairy, often silvery felted; square stems; yellow or white flowers in spikes. Other members of this genus occur in the eastern Mediterranean.

Sideritis canariensis.
Canary endemic; tall shrub; LF locally abundant. Stems and branches felted with yellowish hairs; leaves heart-shaped; well separated whorls of inconspicuous yellow flowers.

Teucrium heterophyllum, **Jocama, Salvia India.**
Macaronesian endemic; small shrub; LO locally frequent. Sage relative; lance-shaped toothed hairy leaves with different leaf forms on a single plant; pink flowers with very long stamens.

Family **CALLITRICHACEAE** - 1 species on Tenerife

Family **PLANTAGINACEAE - plantains etc**. - 13 species on Tenerife (2 Canary endemics)

Plantago webbii, **Crespa.**
Canary endemic; shrublet; HM locally frequent. Linear silky leaves crowded towards tips of stems and pressed towards them; short compact heads of small 4-lobed papery flowers.

Family **MYOPORACEAE** - 1 species on Tenerife

Family **SCROPHULARIACEAE - figworts, foxgloves** - 31 species on Tenerife (6 Canary endemics)

Campylanthus salsoloides, **Romero Marino.**
Canary endemic; small shrub; CO, locally frequent. Alternate linear succulent leaves; pink 5-petalled flowers near tips of stems.

Isoplexis canariensis, **Cresta de Gallo, Bellas de Risco.**
Canary endemic species in a Macaronesian endemic genus; small shrub; LF locally frequent. A very dramatic plant of the laurel forest understory; related to the foxglove; alternate long toothed leaves; erect flowering stems with brilliant orange flowers.

Kickxia sagittata (K.urbanii).
Not endemic; herbaceous perennial; LO local and scarce. A figwort relative which, at first sight, could be taken for a legume; grows in compact clumps; longish simple leaves; solitary yellow spurred flowers. Also occurs in Africa.

Kickxia scoparia.
Canary endemic; small shrub; LO locally abundant. As above but with erect growth; longish simple leaves; yellow spurred flowers.

Scrophularia glabrata, **Hierba de la Cumbre, Fistulera,** a figwort.
Canary endemic; shrublet; HM locally abundant. Square stem; broad leaves with toothed margins; small 2-lipped dark-purple flowers in spikes.

Scrophularia smithii (S.langeana), **a figwort.**
Canary endemic species; woody-based herbaceous perennial; LF locally frequent. Stem and both sides of leaves very hairy; small two-lipped purple to brownish flowers in spikes.

Family **GLOBULARIACEAE - globularias etc**. - 1 species on Tenerife
Globularia salicina.
Macaronesian endemic; small shrub; LO,LF locally frequent. Alternate lance-shaped leaves; globular clusters of white flowers in leaf axils.

Family **OROBANCHACEAE - broomrapes etc**. - 7 species on Tenerife
(1 Canary endemic)

Family **ACANTHACEAE** - 2 species on Tenerife (1 Canary endemic)
Justicia hyssopifolia.
Canary endemic; small shrub; LO locally frequent, Opposite lance-shaped leaves; creamy white 2-lipped hooded flowers; light brown fruit capsules.

Family **CAMPANULACEAE - bellflowers etc**. - 6 species on Tenerife
(1 Canary endemic)
Canarina canariensis, **Bicácaro, Canary Bellflower.**
Canary endemic; scrambling herb; LF locally frequent. Long triangular leaves with strongly toothed margins; large orange bell-shaped flowers with 5 joined petals; edible fruit. Closest relatives on African mountains.

Family **LOBELIACEAE - lobelias** - 1 species on Tenerife

Family **RUBIACEAE - bedstraws, gardenias, coffees** - 15 species on Tenerife (2 Canary endemics)

Phyllis nobla (Phyllis nobila), **Capitana.**

Macaronesian endemic species in a Macaronesian endemic genus; shrublet; LF locally abundant. Oval hairy leaves; 5-lobed whitish flowers in loose clusters; dry black fruit. The genus has its closest relatives in southern Africa.

Phyllis viscosa.

Canary endemic species in a Macaronesian endemic genus; shrublet; LO locally frequent. Narrow sticky leaves; 5-lobed whitish flowers in short dense clusters; dry black fruit.

Plocama pendula, **Balo.**

Canary endemic species in a Canary endemic genus; tall shrub; LO locally abundant. Drooping branches; long narrow leaves; strong smell; very tiny flowers; small round black berries.

Rubia fruticosa, **Tasaigo, a madder.**

Macaronesian endemic; scrambling shrub, LO locally frequent. Found in euphorbia communities; small broad leaves in whorls, prickly below and on margins (which helps distinguish it from *Rhamnus* and *Maytenus*); small 5-lobed pale yellow flowers; berry black or whitish.

Family **CAPRIFOLIACEAE - honeysuckles, elders** - 3 species on Tenerife (1 Canary endemic)

Lonicera etrusca, **Madreselva, a honeysuckle.**

Not endemic; climber; LF locally frequent.Upper pairs of leaves fused at base; large creamy and red 2-lipped flowers in stalked clusters. Also occurs elsewhere in Macaronesia and in the Mediterranean.

Sambucus palmensis, **Saúco, an elder.**

Canary endemic; tall shrub; LF (Anaga) local and scarce. Pithy stem; pinnate leaf with 3 pairs of opposite leaflets and one terminal one, with serrated edges; large flat cluster of small white flowers; small inconspicuous blackish fruit.

Viburnum tinus subsp. *rigidum*, **Follao.**

Canary endemic subspecies; tall shrub, LF widespread and abundant. Broad somewhat hairy opposite leaves; flat clusters of small 5-lobed bell-shaped white flowers; small bluish-black shiny fruit. Forms part of the understory in the laurel forest. The species also occurs in the Mediterranean.

Family **VALERIANACEAE - valerians etc** - 6 species on Tenerife

Family **DIPSACACEAE - scabious, teasels** - 5 species on Tenerife (3 Canary endemics)

Pterocephalus dumetorum, ***Rosalito Salvaje.***

Canary endemic; small shrub; LO locally frequent. A scabious relative; lance-shaped leaves crowded near tip of stem; branched flower head with clusters of 5-lobed pink scabious flowers.

Pterocephalus lasiospermus, ***Hierba Conejera.***

Canary endemic (Tenerife only); small shrub; HM locally abundant. Principal population in Las Cañadas. Cushion-like shrub; long narrow leaves; long-stemmed unbranched flower stem with single clusters of 5-lobed pink scabious flowers.

Family **COMPOSITAE (ASTERACEAE) - daisies etc.** - 171 species on Tenerife (60 Canary endemics).

Ageratina adenophora, ***Azucarera.***

Introduced; herbaceous perennial; LO,LF locally abundant. Broad limp leaves; dense clusters of small white flowers. A common weed, often growing densely by roadsides and in *barrancos*. Native of Mexico.

Allagopappus dichotomus, ***Mato Risco.***

Canary endemic; small shrub; LO locally abundant. Narrow sticky leaves; dense clusters of small yellow thistle-like flowers.

Andryala pinnatifida, ***Estornudera.***

Not endemic; herbaceous perennial; all zones, widespread and frequent. Very variable leaves; clusters of large flat yellow dandelion-like flowers.

Argyranthemum spp., ***Magarza, Margarita.***

Members of a Macaronesian endemic genus with closest relatives in South Africa. 20 species in Canaries (highest diversity of any endemic genus); 9 species on Tenerife, and many subspecies and hybrids; shrublets and small shrubs; all zones, locally frequent. White daisy, with yellow centre.

Argyranthemum coronopifolium/frutescens.

A hybrid (see *TENO* BEYOND *BUENAVISTA*, EXCURSION No.4); small shrub; LO locally frequent. Oblong fleshy jagged leaves; clusters of yellow-centred white daisy flowers.

Argyranthemum teneriffae, ***Margarita del Teide.***

Canary endemic (Tenerife only); shrublet; HM locally abundant. Oblong jagged leaves; clusters of yellow-centred white daisy flowers. One of the commonest plants of Las Cañadas.

Artemisia thuscula, ***Incienso*, a wormwood.**

Canary endemic; small shrub; LO locally frequent. Aromatic plant;

153

silvery divided leaves with flat lobes; golden-brown flowers in dense heads.

Carlina salicifolia, **Cabezote, Cardo de Cristo, a thistle.**
Macaronesian endemic; small shrub; LO,PF locally frequent. Lance-shaped leaves with spiny margins; large yellow flowers.

Carlina xeranthemoides, **Malpica, a thistle.**
Canary endemic; shrublet; HM locally frequent. Very narrow leaves with woolly surface and spiny margins; bright yellow flowers, smaller than in C.*salicifolia.*

Cheirolophus canariensis (Centaurea canariensis), **Cabezón, a knapweed.**
Canary endemic (Tenerife only); small shrub; LO local and scarce. Very jagged leaves; mauve thistle-like flowers.

Cheirolophus teydis (Centaurea arguta), **Cabezón, Cabezuela, a knapweed**.
Canary endemic (Tenerife only); small shrub; HM locally frequent. Lance-shaped toothed sticky leaves; pale yellow thistle-like flowers.

Dittrichia viscosa, **Altabaca.**
Not endemic; small shrub; LO locally abundant. A sticky plant; linear to lance-shaped leaves; yellow flowers. Also occurs in southern Europe.

Gnaphalium spp., **cudweeds.**
Herbs; locally frequent. White felted stems and leaves; dense terminal flower heads with yellowish flowers. Two or three species on the island, one occurring high on Mt. Teide.

Kleinia neriifolia (Senecio kleinia), **Berode, Berol.**
Canary endemic; tall succulent shrub; LO widespread and frequent. Thick-stemmed much branched succulent shrub confusingly like a *Euphorbia* but actually a groundsel relative; terminal lance-shaped fleshy leaves; slender flower heads with pale yellow daisy-like flowers.

Launaea arborescens, **Aulaga, Ahulaga.**
Not endemic; small shrub; LO locally frequent.Dense greyish-green much branched shrub; stems very spiny, giving lattice-work appearance; a few basal leaves but most of the shrub leafless; small solitary yellow flowers. A desert-adapted shrub common in arid areas in the south. Also occurs elsewhere in Macaronesia and in southeast Spain and northwest Africa.

Pericallis cruenta (Senecio cruentus), **Mato Blanco.**
Canary endemic; small shrub; LF locally frequent. Broad ace-of-spades-shaped leaves, slightly jagged, glossy green above and white and woolly below; white daisy-like flowers with yellow centres.

Pericallis lanata (Senecio heritieri), **Palomera.**
Macaronesian endemic; herbaceous perennial; LO locally frequent.

Densely silvery hairy stems and leaves; rounded slightly jagged leaves; showy purplish daisy-like flower with darker centres.

Schizogyne sericea, **Salado, Dama.**

Macaronesian endemic; small shrub; CO,LO locally frequent. Salt tolerant (halophyte); silvery-grey flat linear leaves, sometimes fleshy; clusters of small yellow thistle-like flowers. Closest relatives are found in eastern Africa.

Sonchus spp., **Cerrajas, sow-thistles.**

11 species on the island and some hybrids; small shrubs and herbaceous perennials; LO,LF widespread and frequent. Toothed or jagged leaves; clusters of yellow dandelion-like flowers. Species are hard to distinguish; this genus has undergone adaptive radiation in the Canaries, giving rise to 17 endemic species, as well as some subspecies and hybrids.

Tolpis spp.

6 species on the island, all but one endemic to Tenerife or the Canaries; woody-based herbaceous perennials. Dainty thin stemmed and much branched plant; basal rosettes of jagged leaves; small yellow dandelion-like flowers.

Tolpis webbii, **Flor de Malpaís.**

Canary endemic; woody-based herbaceous perennial; HM locally frequent. As above; first branching occurs at ground level.

Vieraea laevigata, **Amargosa.**

Canary endemic species (Tenerife only); the sole representative of a genus found only in Tenerife; small shrub; LO local and scarce. Fairly broad fleshy leaves; large solitary yellow daisy-like flowers. A fermented liquor used to be made from this plant.

Family **HYDROCHARITACEAE - frog's bits** - 2 species on Tenerife

Family **POTAMOGETONACEAE - pondweeds** - 3 species on Tenerife

Family **ZANNICHELLIACEAE - horned pondweeds** - 1 species on Tenerife

Family **COMMELINACEAE - spiderworts** - 3 species on Tenerife

Family **GRAMINEAE (POACEAE) - grasses** - 129 species on Tenerife (6 Canary endemics)

Arrhenatherum calderae, **Cerrillo de las Cañadas.**

Canary endemic (Tenerife only); grass; HM locally frequent. Tall graceful grass growing in clumps.

Arundo donax, **Caña**, **Giant Reed, Cane.**

Introduced; giant grass; LO locally abundant. A tall bamboo-like cane growing in dense clumps in the bottoms of valleys. The largest 'grass' in Europe, widespread in the Mediterranean though probably originating in Asia. Found throughout Macaronesia.

Vulpia bromoides, **Barren Fescue Grass.**

Not endemic; small annual grass; locally frequent. Can be found near the summit of Mt. Teide. A widespread species.

Poa annua, **Annual Meadow Grass.**

Not endemic; grass; widespread. Can be found near the summit of Mt. Teide. Cosmopolitan.

Family **JUNCACEAE - rushes** - 8 species on Tenerife (1 Canary endemic)

Family **CYPERACEAE - reeds, sedges** - 19 species on Tenerife (4 Canary endemics)

Cyperus capitatus.

Not endemic; sedge; LO locally frequent. A medium sized sedge. Also occurs elsewhere in Macaronesia and the Mediterranean.

Family **TYPHACEAE - bullrushes etc** - 1 species on Tenerife

Family **MUSACEAE - bananas etc** - 1 species on Tenerife

Family **CANNACEAE** - 1 species on Tenerife

Family **PALMAE (ARECACEAE) - palms** - 2 species on Tenerife (1 Canary endemic)

*Phoenix canariensis, **Palma, Palmera,** Canary Palm.*

Canary endemic; tree; LO local and scarce. Possibly no wild trees left, but frequently found in cultivated areas and in towns. Very long pinnately divided leaves. It is closely related to the Date Palm *Phoenix dactylifera,* an introduced species which can be found in some places. The Canary Palm has a thicker trunk and more luxuriant foliage. The fruit are not used for food but the fronds are often used as animal fodder and on the island of La Gomera a type of sweet syrup called *Miel de Palma* is prepared from the sap.

Family **LEMNACEAE - duckweeds** - 2 species on Tenerife

Family **ARACEAE - arums etc**. - 5 species on Tenerife (1 Canary endemic)

*Dracunculus canariensis, **Tacarontilla, Taraguntía.***

Canary endemic; herbaceous perennial; LO,LF locally frequent. Complex palmate leaf; large Lords-and-ladies with flower partially enclosed by conspicuous long bract; reddish-orange berries.

Family **IRIDACEAE - irises** - 6 species on Tenerife

Family **LILIACEAE - lilies etc**. - 39 species on Tenerife (6 Canary endemics)

Asparagus spp., ***Esparrago, Esparraguera*, Asparagus.**

Some endemic species; scrambling shrubs; CO,LO,LF,PF locally frequent. Eight species on the island; hard to distinguish. Needle-like "leaves" in clusters; small whitish bell-shaped flowers in clusters at base of branches; red berries.

Asphodelus spp., **Asphodel.**

Not endemic; herbaceous perennial; LO,PF locally frequent. Three species on the island; strap-like leaves; tall branched flowering stems which remain conspicuous long after the flowers and leaves are dead.

Scilla haemorrhoidalis, **a squill.**

Canary endemic; herbaceous perennial; LO locally frequent. Like *S.latifolia*, but much smaller.

Scilla latifolia, **a squill.**

Not endemic; herbaceous perennial; LO locally frequent. Large bulb giving rise to rosette of strap-like leaves; robust unbranched flowering stem up to 0.5m high with many long stalked, 6-petalled, lilac coloured, star-like flowers. Also occurs elsewhere in Macaronesia (Salvage Islands) and northwest Africa.

Semele androgyna, ***Gibalbera, Gilbarbera*.**

Macaronesian endemic; climber; LF locally frequent. Alternate longish pointed cladodes (look like leaves); small greenish flowers on margins of cladodes.

Smilax canariensis, ***Zarzaparrilla*.**

Macaronesian endemic; climber; LF locally frequent. Thorns on stem; leaves with smooth edge; flat heads of small flowers.

Family **AMARYLLIDACEAE - daffodils etc**. - 4 species on Tenerife (1 Canary endemic)

Pancratium canariense.

Canary endemic; herbaceous perennial; LO locally frequent. Daffodil-like leaves and flowering stem; a terminal cluster of scented white trumpet-shaped flowers.

Family **AGAVACEAE - agaves etc**. - 4 species on Tenerife

Agave americana, ***Pita*, Sisal.**

Introduced; large succulent short-lived perennial; LO locally abundant. A very dramatic plant with enormous rosettes of broad fleshy tapering

157

leaves, each ending in a sharp thorn; the single flower stems can grow up to 10m (33ft) high; numerous very large flat heads of bright yellow blossoms. The plant lives for 10-15 years and then dies after flowering. A native of Mexico, this plant was apparently originally introduced for forage; a very invasive plant which is spreading, especially in disturbed areas, to the detriment of the local vegetation; common on roadsides.

Dracaena draco, **Drago,** **Dragon Tree.**

Macaronesian endemic; tree; LO local and scarce. Tall fat undivided trunk in young trees; older trees divide at the top and these branches may later divide again. The large sword-shaped leaves cluster at the ends of the branches (or at the top of the trunk in the case of younger trees); flat heads of greenish-white flowers; orange fruits. Only a few left growing in the wild state, but the tree is much planted in parks and gardens; several towns, such as Icod, are proud of their very large old trees. Closest relatives around Red Sea and Gulf of Aden.

Family **DIOSCOREACEAE - yams etc**. - 1 species on Tenerife

Tamus edulis, **Norsa.**

Macaronesian endemic; herbaceous perennial climber; LO,LF locally frequent. Alternate broad pointed heart-shaped leaves, flowers in a spike in axil of leaf; reddish-orange berries; related to Yam.

Family **ORCHIDACEAE - orchids** - 7 species on Tenerife (1 Canary endemic)

FERNS

Hansen and Sunding in FLORA OF MACARONESIA list 49 species of ferns and clubmosses, in 18 families, for Tenerife; only one is a Canary endemic. We have restricted the list here to just four conspicuous ferns that are mentioned in the text.

Family **ADIANTACEAE - maidenhair ferns** - 3 species on Tenerife

Adiantum capillus-veneris, **Culantrillo,** **Maidenhair Fern.**

Not endemic; LO local and frequent in damp places. A delicate fern with wiry branches and small fan-shaped "leaflets". Also occurs elsewhere in Macaronesia, the Mediterranean and northwest Africa.

Family **ASPIDIACEAE - buckler ferns** - 4 species on Tenerife

Dryopteris oligodonta, **Helecho macho,** **a buckler fern.**

Canary endemic; LF widespread and frequent. Large fern with compound pinnate leaves.

Family **BLECHNACEAE** - 2 species on Tenerife
Woodwardia radicans, ***Pijara.***
> Not endemic; LF locally frequent. Magnificent 1-2m (3-6ft) fronds.
> Occurs in western Mediterranean.

Family **HYPOLEPIDACEAE - Bracken** - 1 species on Tenerife
Pteridium aquilinum, ***Helecho Hembra*, Bracken.**
> Worldwide; LO,LF,PF locally frequent.

LICHENS

Several hundred species occur on the island; they are hard to identify and
often grow in mixed species clumps. One conspicuous kind is *Usnea* sp.,
which is a long green-grey hanging lichen, frequently seen on older trees in
the moist cloud zone. Another of interest is *Roccella* sp., which grows in
rocky places and was the source of the red and violet dye *Ochil*. *Stereo-
ocaulon* is the grey tufted lichen typically found growing in lava fields.

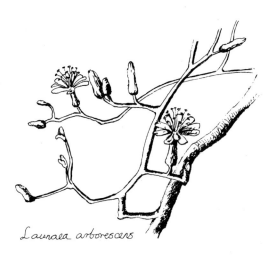

Launaea arborescens

MAMMALS

Tenerife, like most oceanic islands, has very few native mammals; most of the land mammals have been introduced either accidentally or intentionally by humans. This list includes the native, introduced, feral and agricultural mammals that have been recorded from the island and also some of the marine mammals that occur in the surrounding waters. The nomenclature generally follows that of THE ENCYCLOPAEDIA OF MAMMALS, edited by David Macdonald.

CARNIVORA - carnivores

Cat, *Felis catus,* **Gato.**

Feral; cats are found in *barrancos* in the lower zone, in the the pine forest, laurel forest and in the high mountain zone; being largely nocturnal, they are rarely seen. Large specimens, the size of the Wild Cat *(Felis sylvestris)* have been reported from the laurel forest.

Dog, *Canis familiaris,* **Perro.**

Feral; unfortunately stray or abandoned dogs are quite a common sight, especially in the high mountain zone; a few good rabbiters manage to survive the extreme conditions up here and breed. Their offspring are therefore truly wild and can be savage if cornered.

Ferret, *Mustela putorius furo,* **Hurón.**

Feral; much used for hunting rabbits and sometimes escape and live wild.

PERISSODACTYLA - odd-toed ungulates

Donkey, *Burro.*

Agricultural.

Horse, *Caballo.*

Agricultural; mules are still used for transport in some remote parts of the Anaga peninsula.

ARTIODACTYLA - even-toed ungulates

Pig, *Cerdo.*

Agricultural; most pigs are kept intensively in sheds. The island breed of black pig is now very rare indeed although attempts are being made to save it from extinction.

One-humped Camel (Dromedary, Arabian Camel), *Camelus dromedarius,* **Camello.**

Agricultural; a native of southwest Asia and northern Africa, but has never lived "wild" in Tenerife. There are now very few on the island, apart from those used to carry tourists.

Cow, *Vaca.*

Agricultural; there are relatively few dairy cattle. Both cows and bulls are used for pulling ploughs in some areas. No distinct local breed.

Goat, *Cabra.*

Agricultural; very common throughout the island but grazing is forbidden in much of the forest and in Las Cañadas. The ancient Canary breed is now virtually extinct, surviving only on the Desertas Islands (Madeira) where some were introduced by explorers in the 15th century. Recent active interest in the island's heritage has led to plans to bring some of these back to Tenerife.

Sheep, *Oveja.*

Quite large flocks of dairy sheep are kept around La Laguna and La Esperanza; no distinct local breed.

Mouflon, *Ovis musimon,* *Muflón.*

Introduced; originally from the eastern Mediterranean region and Sardinia and Corsica, but widely introduced into other parts of Europe. The smallest of the wild sheep - about 75cm (30in) high; wool dark brown with a lighter coloured "saddle", distinct pale rump patch and short dark tail. Thirteen of these sheep were introduced to the mountains of Tenerife in 1971, amid considerable controversy, to provide additional sport for hunters. They have been hunted since 1977, and although estimates vary it seems that there are well over 100 now. They range down into the forest and up into Las Cañadas, where they have unhindered access to areas with some of the rarest plants in the world - and this in a national park! Their potential for damage is considerable, both from grazing and from spreading alien plant species via seeds that survive in their droppings.

RODENTIA - rodents

Roof Rat (Black Rat), R*attus rattus,* ***Rata de Campo, Rata Negra.***

Introduced; originally from southeast Asia and India, but now a worldwide species. Hard to distinguish from the **Common Rat**; it has relatively larger eyes and ears, and a longer, thinner, uniformly coloured tail; its ears are thinner and almost hairless. Found throughout the island but probably not in villages and towns.

Common Rat (Norway Rat, Brown Rat), *Rattus norvegicus,* ***Rata Común, Rata Gris.***

Introduced; originally from Caspian Sea area but now found worldwide. Hard to distinguish from **Roof Rat**; eyes and ears are relatively smaller, and the ears are finely furred; the tail is shorter and thicker and usually dark above and pale beneath. Found mainly in villages and towns.

Skulls of Giant Rat and Roof Rat

[**Giant Rat**, *Canariomys bravoi*.

This rat is now extinct, but survived until fairly recently; bones (only sub-fossil) are still sometimes found in caves. It was apparently larger than any other European or African members of the family Muridae (rats and mice) and about the size of a rabbit.]

House Mouse, *Mus musculus*, **Ratón**.

Originally from the Turkestan region, but now a worldwide species. This is the only small rodent present on the island, and occurs in all zones.

LAGOMORPHA - rabbits and hares

Rabbit, *Oryctolagus cuniculus*, **Conejo**.

Introduced; originated in the Iberian peninsula and northwest Africa, but introduced to the rest of Europe and much of the rest of the world around 2000 years ago. Rabbits are numerous throughout the island and droppings can be seen even in some volcanic craters and in the middle of lava flows where there is no vegetation! It is thought that they were introduced at the time of the conquistadors (in the 15th century). Rabbit hunting with dogs and guns is a very popular sport, and rabbits are an esteemed food. It seems likely that they have a major influence on the abundance and distribution of some of the native plants; a research program has recently been set up to study this.

INSECTIVORA - insectivores

Algerian Hedgehog, *Erinaceus algirus (Aethechinus algirus)*, **Erizo Moruno**.

Introduced; this species occurs in northwest Africa and southwest

Europe, and also on Gran Canaria, Fuerteventura and Lanzarote. Judging from the number killed on the roads, it is quite common. It is thought that a pair from Morocco were introduced into Fuerteventura in 1892, and some were introduced into Tenerife in about 1903. They are found mainly in the lower zone, but occasionally in the forests and high mountain zone.

White-toothed Shrew, *Crocidura russula,* **Musaraña.**

Not endemic and probably introduced to Tenerife; this species occurs in northern Africa and south-central Europe and is also found on Fuerteventura and Lanzarote. Discovered near Santa Cruz only a few years ago.

Pigmy White-toothed Shrew, *Suncus etruscus,* **Musarañita.**

Not endemic and probably introduced to Tenerife; this species also occurs in Morocco, southwest Europe and parts of Asia. It was discovered in Tenerife only in 1983. This is the smallest living terrestrial mammal; its length is 3.5-4.8cm (1.4-1.9in) and it weighs 2g (0.07oz).

CHIROPTERA - bats

Bats are now rare on the island, having decreased notably in the last few years, probably due to the use of insecticides; they are still seen in some of the remoter *barranco*s and in the high mountain zone. The three species recorded from Tenerife all belong to the family Vespertilionidae.

Canary Long-eared Bat, *Plecotus teneriffae (P.austriacus teneriffae),* **Murciélago.**

Canary endemic species. (Some writers consider it to be a subspecies of a species which occurs in Spain, Atlantic islands, northern Africa and eastward to China and which also reaches southern Britain.) Recorded from the pine forest and high mountain zone.

Western Barbastelle, *Barbastella barbastellus.*

Not endemic; this species occurs in parts of Europe, including Britain, and northern Africa. It is recorded from the laurel forest.

Savi's Pipistrelle, *Pipistrellus savii.*

Not endemic; occurs in southwest Europe, northern Africa, much of Asia, and also on the Cape Verde Islands and other Canary Islands. It is probably the most widespread of the three bats on Tenerife and likely to be the one seen in the lower zone.

PINNIPEDIA - seals

[**Mediterranean Monk Seal,** *Monachus monachus.*

Large colonies used to exist about a century ago, at least on the eastern islands. It still survives in the Mediterranean and in small numbers on Madeira and the westernmost point of Africa.]

CETACEA - whales and dolphins

Many species of whales and dolphins are known to occur in the ocean around the Canary Islands. Most of these prefer deep water and are seldom or never seen close to land, except when found stranded and dying on the shore. In *FAUNA DEL ARCHIPIELAGO CANARIO*, edited by Juan José Bacallado, eight species are mentioned as having been recorded around the islands. We give descriptions of three of the commonest and just list the other five. These three species are all widespread, occurring in all tropical, subtropical and temperate seas. For further information we recommend WHALES OF THE WORLD (1981) by Lyall Watson.

Striped Dolphin, *Stenella coeruleoalba*, **Delfín Listado.**

A small dolphin growing up to 3m (10ft) long with a tall, curved, centrally placed dorsal fin; variable in colour from dark grey or brown to bluish grey on back; lighter grey flanks and white belly; a distinct dark stripe running from the eye along the side of the body and down to the anus. Often more easily seen - and a good diagnostic feature - is a dark wedge which runs down from behind the dorsal fin and forward along the pale flank finishing in a point about half way to the eye. They are usually seen in schools, and often several will be jumping out of the water at a time.

Common Dolphin (Saddleback Dolphin), *Delphinus delphis*, **Delfín Común.**

The smallest dolphin in these waters, growing up to 2.6m (8.5ft) long. Deep grey above and lighter below but with very varied patterns on the sides with patches of deep and light grey and yellow; this criss-cross of patterns is the best way to distinguish them from **Striped Dolphin**. They occur in very active schools with individuals frequently leaping, and often accompany ships.

Bottle-nosed Dolphin, *Tursiops truncatus*, **Delfín Mular.**

A medium-size dolphin growing up to 4.2m (14ft) long; larger and heavier than **Striped Dolphin** or **Common Dolphin**, with a tall, bulky dorsal fin and unpatterned body; often with white patch on tip of lower jaw. Occur in coastal waters and frequently accompany ships. They do not often leap when undisturbed.

Killer Whale, *Orcinus orca*, **Orca Común.**

Goosebeak Whale (Cuvier's Whale), *Ziphius cavirostris*, **Zifio Común, Ballena de Cuvier.**

Great Sperm Whale, *Physeter macrocephalus (P.catodon)*, **Cachalote Común.**

Pygmy Sperm Whale, *Kogia breviceps*, **Cachalote Pigmeo.**

Long-finned Pilot Whale, *Globicephala melaena*, **Calderón.**

BIRDS

This list includes only those birds that breed (or have bred in the recent past) on Tenerife; many more species pass through on migration, and some spend the winter here. About fifty nine species have been recorded breeding on the island in the last few decades; a few of these, however, now have very low populations and the **Red Kite** is extinct. As is usual on islands, the number in any one habitat is lower than would be found in a comparable mainland habitat. The Canary Islands are on the southern extremity of the Palaearctic zoogeographic region (which includes Europe, northern Asia and the Mediterranean part of northern Africa). The breeding birds nearly all belong to the Palaearctic avifauna, rather than the more southern Ethiopian avifauna (which comprises the rest of Africa). We mention on which of the other six main islands a bird can be found; for simplicity we generally consider the three islets north of Lanzarote as being part of that island.

A few bird species are of very restricted distribution, and are either endemic to the Canary Islands or to Macaronesia (Azores, Madeira, Salvages, Canaries and Cape Verdes). There are four Canary Islands endemic species, of which three - **Blue Chaffinch, White-tailed Laurel Pigeon** and **Bolle's Laurel Pigeon** - occur on Tenerife. (The fourth Canary endemic is the Canary Island Chat *Saxicola dacotiae*.) There are also three Macaronesian endemics - **Berthelot's Pipit** (also occurs on Madeira), **Canary** (also occurs on Madeira and Azores) and **Plain Swift** (also occurs on Madeira). In addition, several birds are endemic at the subspecific level, with close relatives in northern Africa or western Europe.

The notes accompanying the list are intended primarily for visitors from northwest Europe; hence we mention subspecific differences from north-west European subspecies where relevant. For those birds that are likely to be unfamiliar, we give sufficient detail to separate them from other resident birds and a few notes on their geographical distribution.

Comments on distribution of breeding birds on Tenerife are very general and normally only refer to broad zones. Reference to index and then to the bird lists in the various excursions in Section 2 will help the reader to find out where there is a chance of seeing any particular species. The breeding distribution of some species is more restricted than their overall distribution and we do not give details of this. Note that when a bird is listed as occurring throughout the island, this does not include the top of Mt.Teide.

The sequence follows that of K.H. Voous ("Ibis" 1973 and 1977); nomenclature, with a few exceptions, follows BIRDS OF THE WESTERN

PALAEARCTIC (1977-. . . .) edited by Stanley Cramp, for non-passerines and THE BIRDS OF THE PALAEARCTIC FAUNA (1959-65) by Charles Vaurie, for passerines. Alternative names are included when there has been a recent change of name, or there is a difference of opinion. Where more than one Spanish name is given, the first is the official Spanish name and the others are local names. A very helpful and comprehensive book on the breeding birds of Tenerife, *ATLAS DE LAS AVES NIDIFICANTES EN LA ISLA DE TENERIFE*, by Aurelio Martín of La Laguna University, was published in Spanish in 1987; this gives up-to-date information about distribution and breeding biology. THE BIRDS OF THE CANARY ISLANDS, by Francisco Pérez Padrón, includes a useful list of well over 100 non-breeding birds that have been recorded on Tenerife. This information can also be found in BIRDS OF THE ATLANTIC ISLANDS by David A. Bannerman, while the bibliography in the book by Aurelio Martín includes many recent publications. A field guide, such as *GUIA DE LAS AVES DE LAS ISLAS CANARIAS* by José Manuel Moreno, would be useful during the migration seasons.

We are indebted to Aurelio Martín Hidalgo and Keith Emmerson (who is the Canary Island contact for the British Ornithologists' Union) for sharing with us their extensive knowledge of the birds of Tenerife.

PROCELLARIIFORMES - petrels

Bulwer's Petrel, *Bulweria bulwerii (B.b.bulwerii)*, **Petrel de Bulwer, Perrito.**

Not endemic; a widespread warm water species that breeds on all the Macaronesian island groups; also occurs on Hierro, and an islet off Lanzarote. A medium sized, long-winged petrel, sombre brownish-black above and below; tail wedge-shaped but normally held closed and thus appears pointed; call has been likened to distant barking of dog. Breeds in the summer in small colonies, mainly on offshore islets; migrates south for the winter.

Cory's Shearwater, *Calonectris diomedea borealis (Puffinus diomedea borealis)*, **Pardela Cenicienta.**

Not endemic; this subspecies occurs in subtropical east Atlantic and breeds on all the Canary Islands; the species occurs in both temperate and subtropical zones of the northern Atlantic, and also in the Mediterranean. Abundant, especially in coastal region. Large, uniformly grey-brown upperparts without any specially dark cap; pure white underparts; heavy yellow bill; graceful gliding flight close to the water following the contours of the waves. Nests in burrows in *barrancos* and on cliffs, and on offshore islets; often feeds within sight of the shore during the day; breeds in the summer; migrates south in October and November, returning to the island in early March.

Little Shearwater, *Puffinus assimilis baroli,* **Pardela Chica, Tahoce.**

Macaronesian endemic subspecies of a species which has larger populations in the southern oceans; it has also been found on Hierro, Gran Canaria and Lanzarote. Present only in small numbers with breeding sites very little known. Small size (in comparison with other shearwaters in the area), slate-blue-black back and pure white underparts; quick beats of wings and alternate gliding close to the surface of the waves; usually seen singly or in twos. Birds can be seen in the vicinity of the islands throughout most of the year.

Storm Petrel, *Hydrobates pelagicus,* **Paíño Común.**

Not endemic; small numbers; has only recently been found breeding on an offshore islet; it occurs in larger numbers on the islets north of Lanzarote, and also on Hierro. Smaller than other petrels in Canarian seas, with square tail, white rump and white stripe under wing; fluttering flight; may follow ships, feeding on plankton churned up by propeller.

Madeiran Storm-petrel, *Oceanodroma castro (O.c.castro),* **Paíño de Madeira.**

Not endemic; a warm water species found on other Macaronesian islands, and elsewhere in the Atlantic and Pacific. A small colony of this species has recently been found on an offshore islet, but breeding has not been confirmed from any of the other Canary Islands; winter breeder. Larger and longer winged than Storm Petrel, with slightly forked tail.

ACCIPITRIFORMES - broad-winged raptors

[Red Kite, *Milvus milvus milvus,* **Milano Real, Villano.**

Declined drastically in recent years and is now apparently extinct on the Canary Islands. Possible reasons for its decline are the lack of carrion (due to more modern agricultural practices), shooting and the wide use of insecticides during locust invasions.]

Egyptian Vulture, *Neophron percnopterus percnopterus,* **Alimoche, Guirre.**

Not endemic; this subspecies occurs on the two easternmost Canary Islands and on Gran Canaria, and also throughout Africa, south Europe and parts of Asia. Very rare on Tenerife and almost certainly no longer breeding. Was apparently common 40 years ago but probably only a couple of individuals are left now.

Sparrowhawk, *Accipiter nisus granti,* **Gavilán.**

Macaronesian endemic subspecies; also occurs on all the western Canary Islands and Gran Canaria (rare) and on Madeira. Found mainly in laurel forest and mixed pine forest.

Buzzard, *Buteo buteo insularum,* **Ratonero Común, Aguililla.**

Canary endemic subspecies (but recent taxonomic revision assigns the Azores population to this subspecies); occurs on all the Canary Islands. Found throughout much of the lower zone (not in extreme south) and also the forest zones. Slightly smaller than the birds in northern Europe *(B.b.buteo),* with underparts more streaked and less densely barred.

Osprey, *Pandion haliaetus haliaetus (P.haliaetus),* **Aguila Pescadora, Guincho.**

Not endemic; this subspecies occurs on all the Canary Islands except La Palma, and in northern Europe and Asia and a few places in Africa and the Middle East. A few may still breed on the island.

FALCONIFORMES - falcons

Kestrel, *Falco tinnunculus canariensis,* **Cernícolo Vulgar.**

Macaronesian endemic subspecies; occurs also on the three western Canary Islands, Gran Canaria and Madeira (another subspecies occurs on Fuerteventura and Lanzarote). Common in all zones on the island. Smaller than birds in northern Europe *(F.t.tinnunculus);* male has more heavily streaked darker crown; rest of upper parts deeper chestnut with larger spots and underparts deeper cream; female more heavily barred with upper tail-coverts tending to grey-blue.

Peregrine Falcon (Barbary Falcon), *Falco peregrinus (F.peregrinus pelegrinoides, F.pelegrinoides pelegrinoides),* **Halcón de Berbería, Halcón Tagarote.**

Not endemic; this species has an almost worldwide distribution; it also occurs on Gran Canaria and the two easternmost Canary Islands. (Recent studies show that populations sometimes treated as *Falco pelegrinoides* are merely forms of *F.peregrinus;* Canary Island breeding birds have not yet been assigned to a subspecies.) Very rare. There is no definite proof of breeding on the island.

GALLIFORMES - game birds

Barbary Partridge, *Alectoris barbara (A.b.koenigi),* **Perdiz Moruna.**

Not endemic, possibly introduced; occurs in Sardinia, northern Africa and also on Gomera, Lanzarote and Fuerteventura. (Some writers consider the island birds, along with those from northwest Morocco, to belong to a subspecies *A.b.koenigi.*) Frequent in much of the lower zone, southern pine forest and high mountain zone. Similar to Red-legged Partridge *(A.rufa)* but slightly smaller and with chestnut-coloured rather than black chest band.

Quail, *Coturnix coturnix coturnix (C.c.confisa),* **Codorniz.**

Probably not endemic, but taxonomic situation confused. *C.c.coturnix* is the migratory Quail that occurs throughout much of Europe. Some authors, however, have considered that the Quail breeding on the western Canary Islands and on Madeira form a separate, resident subspecies, *C.c.confisa*, but that migratory birds sometimes breed on Gran Canaria and the easternmost Canaries and perhaps also on the other islands. Patchy distribution along the north side of the island.

Tufted Guinea Fowl, *Numida meleagris*, **Gallina de Guinea.**
Introduced; occurs throughout tropical Africa. Recently introduced in the coastal region of Bajamar in the north of the island where there is now a small semi-wild population. A large bird - about the same size as a female Capercaillie *(Tetrao urogallus)* - with conspicuous red crest, naked head and neck and grey plumage with white spots.

GRUIFORMES - rails

Moorhen, *Gallinula chloropus chloropus*, **Polla de Agua, Patito.**
Not endemic; assumed to belong to this widespread subspecies; also occurs on Gran Canaria and Fuerteventura. Restricted to a few ponds near Bajamar.

CHARADRIIFORMES - waders and gulls

Cream Coloured Courser, *Cursorius cursor (C.c.bannermani)*, **Corredor, Engaña Muchachos.**
Not endemic; this species also occurs on Fuerteventura and Lanzarote and formerly bred on Gran Canaria; there are also scattered populations in northern Africa. Although there are no definite recent records of breeding on Tenerife there have been regular sightings in the southern coastal area. It is about the size of a Turnstone *(Arenaria interpres)*, sandy coloured with black and white eye stripe and curved bill. It runs fast when disturbed, and then suddenly stands very still.

Stone Curlew, *Burhinus oedicnemus distinctus*, **Alcaraván, Pedro-Luis.**
Canary endemic subspecies; occurs only on the three western islands and Gran Canaria (another subspecies occurs on Fuerteventura and Lanzarote). Rare and only found in the southern coastal zone. This bird is active mainly at night; it is hard to flush in daytime and squats close to the ground if alarmed.

Little Ringed Plover, *Charadrius dubius curonicus*, **Chorlitejo Chico.**
Not endemic; this subspecies occurs throughout Europe and also on Gran Canaria, Fuerteventura and in northern Africa. Found only in southern coastal area, usually beside freshwater or brackish pools. Distinguished from Ringed Plover *(C.hiaticula)*, which occurs on migration, by the lack of a white wing bar in flight and the presence of

a white line above the black on the forehead. The species was not recorded on Tenerife until the twentieth century.

Kentish Plover, *Charadrius alexandrinus alexandrinus*, *Chorlitejo Patinegro.*

Not endemic; this subspecies occurs in southern Europe and has been found in the past on all the Canary islands, but is probably no longer on the three westernmost islands; the species occurs in coastal areas throughout much of the northern hemisphere. Found only in southern coastal area.

Woodcock, *Scolopax rusticola*, *Chocha Perdiz, Gallinuela.*

Not endemic; also occurs on La Palma and Gomera. Found in the laurel forest, pine and chestnut forest of the north of the island.

Herring Gull, *Larus argentatus atlantis*, *Gaviota argéntea.*

Macaronesian subspecies; occurs on all the Canary Islands and in the Azores and Madeira. A few medium-sized colonies and scattered pairs on sea cliffs and islets. It has yellow legs and a darker mantle than northwest European birds, and thus could be mistaken for the Lesser Black-backed Gull *(Larus fuscus).*

Common Tern, *Sterna hirundo hirundo*, *Charrán Común, Garajao.*

Not endemic; this subspecies occurs throughout northern Europe, and has been recorded nesting on all the Canary islands in the past. Nesting is now very rare on Tenerife, although it is still recorded on Hierro and La Palma.

COLUMBIFORMES - pigeons

Rock Dove, *Columba livia (C.l.canariensis)*, **Paloma Bravía, Paloma Salvaje.**

Not endemic; occurs on all the islands. (The birds from the Canaries, together with those from Madeira, are sometimes considered as a separate subspecies.) Widely distributed throughout the island; the most characteristic bird of deep *barrancos.*

Bolle's Laurel Pigeon (Long-toed Pigeon), *Columba bollii (C.trocaz bollei, C.trocaz)*, **Paloma Turqué, Turcón, Paloma Torquesa, Torcaz.**
Canary endemic species; also occurs on La Palma, Gomera and Hierro. (This species has sometimes been considered as being a Canary Island form of *Columba trocaz*, which occurs on Madeira.) Found in the laurel forest. Like a Woodpigeon *(C.palumbus)*, but lacks the white wing bars and the white patch on the side of the neck; the tail has a pale subterminal band, beyond which is a dark bar. Feeds on the fruit of various laurel species and also **Myrica** and **Rhamnus**.

White-tailed Laurel Pigeon (Laurel Pigeon), *Columba junoniae*, ***Paloma Rabiche, Paloma Rabil, Rabo Blanco.***

Canary Island endemic species (also occurs on La Palma and Gomera). Rare; found only in the laurel forests of the Orotava valley and Teno peninsula. In flight it can be distinguished from **Bolle's Laurel Pigeon** by its browner upperparts and by the tail, which becomes paler towards the tip, without any dark terminal bar. It is a less specialized feeder than **Bolle's Laurel Pigeon** and is especially fond of the fruits of **Ocotea**. It was not 'discovered' in Tenerife until 1975. Although this bird has more often been called simply **Laurel Pigeon** we have preferred to use the name given in BIRDS OF THE ATLANTIC ISLANDS (1963) by David Bannerman, so that the two closely related species can be referred to collectively as laurel pigeons.

Turtle Dove, *Streptopelia turtur turtur*, ***Tórtola Común.***

Not endemic; this subspecies occurs throughout Europe and Turkey and also on all the Canary Islands. Found primarily in the pine forest but also in parts of the laurel forest and lower zone. Summer visitor, arriving in early spring and leaving for Africa in late September or October.

PSITTACIFORMES - parrots

Several species of parrots, presumably escaped from captivity or deliberately set free, now live in the parks of Santa Cruz, La Laguna and Puerto de la Cruz. It is probable that some of them nest there.

STRIGIFORMES - owls

Barn Owl, *Tyto alba alba*, ***Lechuza Común, Coruja.***

Not endemic; this subspecies is the northern European one and also occurs on Gran Canaria, Hierro and perhaps La Palma and La Gomera (another subspecies occurs on Fuerteventura and Lanzarote). Patchy distribution in lower zone, often near habitations. It nests in holes and caves.

Long-eared Owl, *Asio otus canariensis*, ***Búho Chico.***

Canary endemic subspecies; also occurs on all the three western Canary Islands and on Gran Canaria. Widespread throughout the island, although least common in the pine forest. It is smaller than the more northern subspecies *(A.o.otus)*, with more heavily mottled and streaked plumage. It has a broader habitat range than its northwest European counterpart, often being found in open country. It frequently nests on the ground under clumps of ***Cardón***, in rock crevices in *barrancos* and occasionally in the crowns of palm trees. Its diet seems to consist mainly of small birds.

APODIFORMES - swifts

Plain Swift (Little Black Swift), *Apus unicolor (A.u.unicolor),* ***Vencejo Unicolor, Andoriña, Avurrión.***

Macaronesian endemic species; occurs only on Madeira and the Canary Islands. Common throughout Tenerife, especially during the summer months; it is a partial migrant and a large part of the population spends the winter in Africa. It is smaller and darker than the **Pallid Swift** and the throat is hardly paler than the breast; it also flies faster. Its scream has a rapid trill.

Pallid Swift (Pale Swift), *Apus pallidus brehmorum,* ***Vencejo Pálido.***

Not endemic; this subspecies occurs in Madeira, in southern Europe and northern Africa; the range of the species extends further east to Pakistan. Very rare and often confused with other swifts; not certain if it still breeds on Tenerife, and is probably most frequent in the lower zone but numbers have declined in recent years; it migrates to Africa for the period October to January. In contrast to Common Swift *(Apus apus)*, which occurs here on migration, it is browner, less blackish, has paler forehead and more conspicuous white throat and slower wing beats. In contrast to the **Plain Swift** it is larger, has sickle-shaped wings and a more gliding flight.

CORACIIFORMES - hoopoes and allies

Hoopoe, *Upupa epops (U.e.pulchra),* ***Abubilla, Tabobo.***

Not endemic; this species occurs in Europe, Asia and northern Africa and on all the Canary Islands. (Some writers have named one subspecies for the western Canary Islands and Gran Canaria and another for Fuerteventura and Lanzarote.) Patchily distributed in the lower zone, especially in agricultural areas; also occurs in the high mountain zone. A spectacular bird, the size of a Mistle Thrush *(Turdus viscivorus)*, with orange-buff head and body, a conspicuous crest and long curved bill; wings and tail black with bold white bars; feeds on the ground.

PICIFORMES - woodpeckers

Great Spotted Woodpecker, *Dendrocopos major canariensis,* ***Pico Picapinos, Pájaro Carpintero, Pájaro Peto.***

Tenerife endemic subspecies (there is another subspecies on Gran Canaria). Found in the pine forest, especially among mature trees. Very similar to the northern European forms.

PASSERIFORMES - songbirds

Lesser Short-toed Lark, *Calandrella rufescens*, **Terrera Marismeña, Calandria.**

Subspecies situation complex; this species occurs in eastern Europe and south Spain and also on Gran Canaria, Fuerteventura and Lanzarote. On Tenerife, the population around Los Rodeos airport near La Laguna has been described as an endemic subspecies, *C.r.rufescens*, while those found in the southern coastal zone are said to belong to the more widespread subspecies, *C.r.polatzeki*, which is the one represented on some of the other Canary Islands. Smaller than Skylark *(Alauda arvensis)*, with a shorter bill; short musical song, often while flying upward in a spiral.

Berthelot's Pipit, *Anthus berthelotii berthelotii*, **Bisbita Caminero, Caminero, Correcaminos.**

Canary endemic subspecies of a Macaronesian endemic species; it occurs on all the Canary Islands (another subspecies occurs on Madeira). Very common throughout the island, but especially in uncultivated parts of the lower zone and in the high mountain zone. About the same size but greyer than Meadow Pipit *(A.pratensis)*; it sometimes sings in the air.

Grey Wagtail, *Motacilla cinerea canariensis*, **Lavandera Cascadeña, Alpispa.**

Canary endemic subspecies; also occurs on La Palma, Gomera and Gran Canaria. Common by running water in lower and forest zones. Similar to the birds in northwest Europe, but distinctly darker and richer yellow below. In contrast to northwest Europe it often occurs in towns and gardens on Tenerife.

Robin, *Erithacus rubecula superbus*, **Petirrojo, Papito.**

Canary endemic subspecies; also occurs on Gran Canaria (another subspecies occurs on Hierro, La Palma, Gomera, other Macaronesian islands and elsewhere. Found throughout most of the forest zone and in upper parts of the lower zone; not a common garden bird. Darker and richer red than the subspecies found in Britain, with a whiter belly; the song is more rich and varied and has phrases reminiscent of the Song Thrush *(Turdus philomelos)*.

Blackbird, *Turdus merula cabrerae*, **Mirlo Común.**

Macaronesian endemic subspecies; also occurs on the three western Canary Islands, Gran Canaria and on Madeira (some authors have considered the birds on Hierro and La Palma to be a separate subspecies,

T.m.agnetae). Found throughout the forest zone and in much of the lower zone. Smaller than birds in northwest Europe; the male is deeper and glossier black, and the female more blackish brown; song slightly different.

Spectacled Warbler, *Sylvia conspicillata orbitalis,* **Curruca Tomillera, Zarzalero, Chirrera.**

Macaronesian endemic subspecies of a species which occurs in northern Africa and the Mediterranean region; the subspecies occurs on all the Canary Islands, Madeira and the Cape Verde Islands. (A recent taxonomic revision treats the Madeiran birds as a separate subspecies *S.c.bella*, and considers the Canary population as intermediate between the two subspecies.) Widespread in the lower zone and high mountain zone, and also in the southern pine forest. It is similar to the Whitethroat *(Sylvia communis)*, but the male has a darker head (especially forehead) and the female is greyer; both sexes have a conspicuous chestnut wing patch, white outer tail feathers, pale eye-ring and yellowish legs; the song is also reminiscent of the Whitethroat.

Sardinian Warbler, *Sylvia melanocephala leucogastra (S.m.melanocephala),* **Curruca Cabecinegra, Capirollo.**

Canary endemic subspecies of a species which occurs in the Mediterranean region and northern Africa; it also occurs on Gran Canaria and the three westernmost islands, while the typical subspecies occurs on the two easternmost islands (although some writers do not consider that the Canary form should be separated from *S.m.melanocephala*). Found throughout the lower zone (not the extreme south) and parts of the laurel forest. The head of the male is jet black (this extending well below the eye) but with the throat white; the whole of the upperparts including the wings are dark grey; the female is somewhat browner, with dark grey head. Both sexes have white outer tail feathers and a conspicuous red ring around the eye.

Blackcap, *Sylvia atricapilla heineken,* **Curruca Capirotada, Capirote.**

Macaronesian endemic subspecies; also occurs on the three western Canary Islands, Gran Canaria and on Madeira. Common throughout much of the lower zone, especially in gardens and cultivated areas, and in the laurel forest; it is also a common cage bird. Both sexes are darker and slightly smaller than the birds of northwest Europe and can be somewhat confusing; however, the lack of white in the outer tail feathers and the black or brown crown readily distinguishes it from the other two resident *Sylvia* warblers. Its song is slightly shorter than that of the European form and it also has a quite different drawn out 'churr'.

Chiffchaff, *Phylloscopus collybita canariensis*, **Mosquitero Común, Hornero, Chivita.**

Canary endemic subspecies; also occurs on the three western islands and Gran Canaria. Common throughout the island except in the extreme south. Found in areas with trees and also in treeless scrub areas. It is darker olive than the birds in northwest Europe and more tawny buff below; the song is quite distinct with considerable individual variation, and at times reminiscent of the declining scale of the Willow Warbler *(P.trochilus)*. In the absence of other leaf-warblers, it is found in a greater range of habitats than in northwest Europe.

Goldcrest (Firecrest), *Regulus regulus teneriffae (Regulus ignicapillus teneriffae)*, **Reyezuelo, Banderita.**

Canary endemic subspecies; also occurs on Hierro, La Palma and Gomera. (Some writers, including a very recent revision, consider this bird to be a Firecrest *R.ignicapillus*.) Found in the laurel forests and pine forests, although rarer in the south. Black sides of the crown somewhat broader than in the subspecies of Goldcrest occurring in northwest Europe, and continuing across the forehead; bill slightly longer and general colouring slightly darker; the song is shorter and harsher.

Blue Tit, *Parus caeruleus teneriffae*, **Herrerillo Común, Chirrero.**

Canary endemic subspecies; also occurs on Gran Canaria and La Gomera (three other island subspecies are recognised: one on Hierro, one on La Palma and one on Fuerteventura and Lanzarote). Common throughout the island except in the southern coastal area. It differs from the continental Blue Tit in having a strikingly dark head, almost no white on the wings and a proportionately longer and thinner beak; the calls are very varied and include some which are deceptively like the Great Tit *(Parus major)*, which does not occur here; other calls are similar to those of other tits. In the absence of other tit species, it is found in a greater range of habitats than in northwest Europe, including the pine forest.

Great Grey Shrike, *Lanius excubitor koenigi*, **Alcaudón Real Moruno.**

Canary endemic subspecies; also occurs on Gran Canaria, Fuerteventura and Lanzarote. Patchy distribution in southern lower zone and high mountain zone. It is paler in colour than the birds in northwest Europe, with slightly shorter tail and wings; it does not migrate.

Raven, *Corvus corax tingitanus*, **Cuervo.**

Not endemic; this subspecies occurs on all the Canary Islands and also in northern Africa. Formerly common in all zones, but now limited to the more remote areas. The plumage is very glossy, the bill shorter and more stumpy than in the birds of northwest Europe; the call is somewhat more strident.

Starling, *Sturnus vulgaris*, **Estorino Pinto.**

Not endemic; it is a fairly common winter visitor and in recent years a few pairs have nested in buildings in La Laguna.

Spanish Sparrow, *Passer hispaniolensis hispaniolensis*, **Gorrión Moruno, Palmero.**

Not endemic; this subspecies occurs on all the Canary Islands, in the western Mediterranean, northern Africa and parts of Asia; another subspecies occurs in much of Asia. Patchy distribution in lower zone and in urban areas. The male has a chestnut crown, and its cheeks are whiter than the House Sparrow; it has a black breast and more dark streaks on flanks and back. It colonized the island in the late nineteenth century, although it was in the eastern islands before this. It takes the place of the House Sparrow *(P.domesticus)*, which does not occur here.

Rock Sparrow, *Petronia petronia madeirensis*, **Gorrión Chillón**.

Macaronesian endemic subspecies of a species that occurs throughout southern Europe and the Mediterranean region; the subspecies also occurs on Hierro, La Palma, Gomera, Gran Canaria and Madeira. (Some authors doubt whether it is distinct from the Mediterranean subspecies *P.p.petronia*.) Patchy distribution in lower zone. It is like a female House Sparrow *(Passer domesticus)*, but with a long pale eye stripe and a long pale stripe on the crown; pale tail markings are visible when it flies; very noisy and usually in flocks.

Chaffinch, *Fringilla coelebs tintillon*, **Pinzón Vulgar, Chau-chau, Pájaro de Monte.**

Canary Island endemic subspecies; also occurs on Gomera and Gran Canaria (another subspecies occurs on Hierro and yet another on La Palma). Common in, and just below, the forest zone along north side of the island; it overlaps with the **Blue Chaffinch** to a limited extent in areas of mixed pine forest (see *EL LAGAR*, EXCURSION No.16). The male differs from the European **Chaffinch** in having slate-blue instead of chestnut mantle, very dark slaty crown and especially forehead, and paler pink breast. It differs from the male **Blue Chaffinch** in its smaller size, but also in its pinkish cheeks and breast, green rump, strong white bars on the wing, yellow edges to the main wing feathers, and white in the tail; all these are suffused with blue in the male **Blue Chaffinch**. The female is greenish-brown above and buff below; she could be confused with the female **Blue Chaffinch**, but is distinctly browner, smaller and with a shorter bill. The song has the same general pattern as in the British subspecies *(F.c.gengleri)*, but individual birds tend to have a greater repertoire.

Blue Chaffinch, *Fringilla teydea teydea, **Pinzón Azul**, **Pájaro Azul del Teide**, **Pájaro de la Cumbre***.

Tenerife endemic subspecies of a Canary endemic species (another subspecies occurring in Gran Canaria). Common in pine forests; it is not found in the lower cultivated zone. It is larger than the **Chaffinch,** the male being rather uniform slate-blue, whitish on the belly and under the tail, and with a white eye ring; it entirely lacks pink on the breast. The female is dull grey all over, with only traces of green and brown; the belly is paler. The song is simpler and with fewer variations than in the **Chaffinch**. The diet consists primarily of pine seeds which it cracks with its sturdy bill; it is more insectivorous during the nesting season.

Serin, *Serinus serinus, **Verdecillo***.

Not endemic; the species occurs throughout southern Europe and in the Mediterranean region. Recently reported locally in lower zone in southeast of the island, and near Puerto de La Cruz and La Laguna; could be the result of pet shop escapes. Smaller than the **Canary** but very similar; male has yellower head and a yellow rump.

Canary, *Serinus canaria (S.canarius), **Canario***.

Macaronesian endemic species; occurs on all the Canary Islands except Fuerteventura and Lanzarote, and also on Madeira and the Azores. Found throughout the island except in the southern coastal area. It is slightly bigger than a **Goldfinch**, with greyish streaked back and yellowish head and underparts - but never bright yellow all over like the caged variety. Its song, however, is like that of the caged bird, but is often delivered in flight.

Greenfinch, *Carduelis chloris aurantiiventris, **Verderón, Verderón Común***.

Not endemic; this subspecies also occurs on Gran Canaria, Madeira, the Iberian peninsula, several islands in the Mediterranean and northern Africa. Patchy distribution in the lower zone and lower forest zone along the north of the island. It was first recorded breeding in Tenerife in 1966.

Goldfinch, *Carduelis carduelis parva, **Jilguero**, **Pájaro Pinto***.

Not endemic; this subspecies occurs on all the islands except Hierro, and also elsewhere in Macaronesia, in the Mediterranean region of Europe and in northern Africa. Patchy distribution in lower zone up to the edge of the laurel forest. It is smaller than birds in northwest Europe and has a more slender bill; it is usually somewhat darker. A popular cage bird.

Linnet, *Carduelis cannabina meadewaldoi (Acanthis cannabina meade-waldoi), **Pardillo Común**, **Millero***.

Canary endemic subspecies; also occurs on Hierro, La Palma, Gomera and Gran Canaria (another subspecies occurs on Fuerteventura and Lanzarote). Found throughout the lower zone and cultivated areas. Darker and more richly coloured than the birds in northwest Europe and with a larger and slightly thicker bill. A popular cage bird.

Trumpeter Finch, *Bucanetes githagineus amantum (Rhodopechys githaginea amantum)*, **Camachuelo Trompetero, Pájaro Moro.**

Canary endemic subspecies of a species which is widespread in the Sahara, Middle-east and eastwards into Asia; it also occurs on La Gomera, Gran Canaria, Fuerteventura and Lanzarote. Local in the southern coastal area. A small finch, only slightly larger than the **Goldfinch**. The male has a stout bright red bill and pinkish-brown plumage in the breeding season; female is duller brownish with yellowish-brown bill.

Corn Bunting, *Miliaria calandra (Emberiza calandra calandra, E.c.thanneri)*, **Triguero.**

Not endemic; occurs on all the islands. (Some writers have considered the Canary form to be a separate subspecies.) Patchy distribution in lower zone, typically in cultivated area. Virtually indistinguishable from continental European birds; its song is said to be slightly softer.

Cory's Shearwater

REPTILES

LIZARDS

Family **LACERTIDAE**

Canary Lizard, *Gallotia galloti (Lacerta galloti)*, **Lagarto.**

Canary endemic species of a Canary endemic genus (sometimes considered to belong to the widespread genus *Lacerta*). Some authorities say there are two subspecies on Tenerife (*galloti* in the south and *eisentrauti* in the north) and another *(insulanagae)* on an offshore rock. (The other western islands each have a separate subspecies of the same species, whilst different endemic species live on the other islands.) **Canary Lizard** is not really a good name for this lizard because of the presence of the other species on the islands; however, we use it because it is used in recent texts on reptiles. The species is widespread throughout the island, especially in the lower and high mountain zones but also in open places in the forests; it can also be found close to the peak of Mt Teide. This is a heavily built lizard showing enormous variation within the island. The adult male can reach 300mm (12in) in length and is dark brown or blackish, more or less spotted with blue and green; the female and young are often rich brown above with pale longitudinal stripes. It is active by day, and although shy can usually be attracted by an applecore or piece of bread or tomato. It is mainly vegetarian, but also eats small animals such as earthworms and snails.

[Giant Lizard.

Extinct; a number of large lizards have been found as fossils and subfossils on Tenerife and have been given a variety of names: *Gallotia* (or *Lacerta*) *goliath, maxima* and *simonyi*. Although the precise relationships of these forms are still under discussion it seems clear that at least one type of lizard considerably larger than the existing **Canary Lizard** was present on Tenerife in the past. The state of some of the remains suggest that these lizards survived until after the arrival of people on Tenerife; small populations of **Giant Lizards** still survive on Hierro and Gran Canaria. (See *CUEVA DE SAN MARCOS*, EXCURSION No.21.)]

Family **SCINCIDAE**

Canary Skink, *Chalcides viridanus viridanus*, **Lisa**.

Canary endemic subspecies of a Canary endemic species; the subspecies also occurs on Hierro (another subspecies occurs on Gomera); the species also occurs on Madeira, but is thought to have been introduced there. **Canary Skink** is not really a good name for this skink because two other species of skinks occur further east in the Canary Islands; however, we use it because it is used in recent texts on reptiles. Widespread, but most common in the lower part of the island, especially in the north; secretive and normally encountered only by turning over rocks or logs. A small lizard, less than 120mm (5in) long, and rather thin and cylindrical, with a small head and short weak legs; it is very glossy, appearing somewhat iridescent, warm brown on the back with rows of pale spots, blackish on the sides.

Family GEKKONIDAE
Canary Gecko, *Tarentola delalandii (T.d.delalandii)*, **Perenquén.**

Canary endemic species (sometimes considered as a Canary endemic subspecies of a Macaronesian endemic species). The same form is found on La Palma, but those on the other western Islands and Gran Canaria are distinct and are treated either as different subspecies or species; a separate species occurs on the eastern islands. According to current practice we call this the **Canary Gecko**, although, as in the case of the other lizards, we do not think it a good name when there are several other species on the Canary Islands. It is found from sea level up to Las Cañadas. It is a flat-bodied lizard with warty skin and large eyes; it can change colour from almost black to very pale quite rapidly; the adhesive pads on the toes make it an excellent climber, even on overhanging surfaces. It is often active at night, and can be seen hunting insects attracted to lights in towns.

Turkish Gecko, *Hemidactylus turcicus.*

Not endemic; occurs in the Mediterranean region. Only recently introduced onto the island and found in the region of Santa Cruz. It is most easily distinguished from the **Canary Gecko** by the colour of its tail, which has alternate rings of pale brown and black, and by the presence of claws on all its toes.

TURTLES AND TORTOISES

Four species of sea turtles are recorded from the waters around the Canary Islands, although none of them breed here. They are hard to identify at sea, but the brief descriptions that we give should help in identification of individuals that become stranded on beaches. There are no living land tortoises, but we include a comment on one fossil species.

Family DERMOCHELYIDAE
Leathery Turtle, *Dermochelys coriacea*, **Tortuga Laúd.**

Widespread species occurring in the Atlantic, Mediterranean, Pacific and Indian Ocean. The largest turtle in these waters, with shell growing up to 180cm (71in) or more. Apart from growing larger than the other turtles, it can be distinguished by the fact that its whole shell is covered with black or dark brown skin, and has 5 to 7 prominent longitudinal ridges.

Family **CHELONIIDAE**
Loggerhead Turtle, *Caretta caretta*, **Tortuga Boba.**
Widespread species occurring in the Atlantic, Mediterranean, Pacific and Indian Ocean. The commonest of the turtles seen around the coast. This is a horny-shelled turtle with shell growing up to about 110cm (43in) long, although it is usually smaller than this; it is oval and rather long; there are 5 costal plates (clearly marked areas on the upper side of the shell) along each side.

Hawksbill Turtle, *Eretmochelys imbricata*, **Tortuga Carey.**
A tropical turtle found in the Atlantic, Pacific and Indian Ocean; only rarely occurs in northern Europe. Sometimes seen near the coast. A horny-shelled turtle with shell growing up to 90cm (35in) long or more; it is easily distinguished from other species by having overlapping horny plates on its shell; there are only 4 costal plates along each side.

Green Turtle, Edible Turtle, *Chelonia mydas*, **Tortuga Verde.**
A species of warm seas; very rarely seen in European waters. A large horny-shelled turtle with shell growing up to 140cm (55in) long; the shell is oval and brown or olive with darker markings; the plates do not overlap; there are only 4 costal plates along each side.

Family **TESTUDINIDAE**
[**Giant Tortoise**, *Geochelone burchardi*.
Extinct; fossil bones of this tortoise have been found in volcanic deposits in a quarry near Adeje in the south of the island, but it evidently became extinct long before the arrival of humans. It was probably over 80cm (31in) long.]

AMPHIBIANS

Family **RANIDAE**
Marsh Frog, *Rana perezi (R.ridibunda perezi)*, **Rana Común, Rano, Sapo.**
Not endemic; this species (which was previously considered a subspecies of *Rana ridibunda*) also occurs on the three westernmost islands, on Gran Canaria and probably also on Lanzarote and Fuerteventura. It was probably introduced to the Canary Islands by the Spanish colonists for food. The species is native to the Iberian peninsula and northwest Africa; some older books call it *Rana esculenta* or *R.e.ridibunda*. It is

abundant wherever fresh water is available, living mainly in irrigation tanks in cultivated areas. An ordinary large frog, reaching a length of 150mm (6in); very variable in colour but usually olive with dark markings. Its voice is very varied and erratic.

Family **HYLIDAE**

Stripeless Tree Frog (Mediterranean Tree Frog), *Hyla meridionalis (H.arborea meridionalis)*, ***Ranita de San Antonio.***

Not endemic; also occurs in southern parts of France, the Iberian peninsula, northwest Africa and Madeira (it was previously considered a subspecies of *Hyla arborea*). It occurs on all the Canary Islands but there is doubt as to whether its presence is natural or due to introduction by humans. Very common in the lower parts of the island. It grows up to 5cm (2in) long; it has smooth skin, long legs and adhesive pads on the tips of the fingers and toes; it is variable in colour but usually green (and occasionally spotted) with a dark mark behind the eye which extends hardly - if at all - beyond the level of the front legs; lives in trees and bushes, often high up and quite far from water; its voice is a deep croak repeated at one second intervals or less. It has been calculated that there might be up to one million individuals per square mile in banana plantations in the Orotava valley.

Stripeless Tree Frogs

FRESHWATER FISH

The **European Eel** is the only naturally occurring species of freshwater fish in Tenerife; four others have been introduced.

Family ANGUILLIDAE - freshwater eels
European Eel, *Anguilla anguilla*, **Anguila.**
   1.5m (5ft). Not endemic; can be found in some *barrancos* in north Tenerife. It also occurs on Gomera and La Palma. These elongate and snake-like fish breed in the Sargasso Sea near the West Indies; the larvae drift across the Atlantic in the Gulf Stream and find their way to fresh water. The young eels (elvers) develop in fresh water until mature at 8-10 years old, and then they return to sea for their long migration back to the spawning grounds. Their colour varies, according to developmental stage: they are silvery when mature.

Family CYPRINIDAE - minnows and carps
Common Carp, *Cyprinus carpio*, **Carpa.**
   1m (3.3ft). Introduced as a decorative fish; is of Asian origin.
Goldfish, *Carassius auratus*, **Carpín Dorado.**
   Introduced as a decorative fish; is of Asian origin.

Family POECILIIDAE - livebearers
Guppy, *Poecilia reticulata*, **Guppy.**
   5cm (2in). This small fish, originally from South America, has been introduced for mosquito control; it can be found in reservoirs and water tanks.
Western Mosquitofish, *Gambusia affinis*, **Gambusia.**
   7.6cm (3in). Another small fish, originally from North America, which has also been introduced for mosquito control.

SELECTED MARINE FISH

Marine fish occupy a wide range of habitats. They are either free-swimming (pelagic) or bottom-living (demersal); the free swimming ones may stay close to the shore (coastal pelagic), live around rocks or live entirely in the high seas (oceanic pelagic) either near the surface or in deep water. The bottom-living species may prefer a hard rocky sea-bed or a pebbly, sandy or muddy one; they may live close to the shore or further out on the continental shelf. Some fish are solitary or occur in very small groups, whilst others live in shoals, and in many cases are migratory.

Over 300 species of fish have been recorded from the waters around Tenerife; these belong to at least 80 different fish families. We list about 180 species which are common or fairly common, although we include notes on only about one third of these. Several additional species are found around the eastern islands of Lanzarote and Fuerteventura, and yet others around the western islands of Hierro and La Palma. For people who are interested in furthur information we suggest two books recently published by Gobierno de Canarias (see APPENDIX).

In selecting the species for comment we have used three main criteria. First, we have concentrated on species that are common or fairly common. Second, we have selected those that can be found close inshore over the continental shelf at less than a depth of 100m: this includes species that you might see in rock pools, when snorkelling, or out in a boat close to the shore, and also those that are caught by rod off rocky coasts. Third, we indicate the fish which are good to eat and might be encountered in the fish market and restaurants. Marine fish are not included in the index, apart from a few mentioned in Sections 1 and 2.

The order of families is that used in FISHES OF THE NORTH-EASTERN ATLANTIC AND THE MEDITERRANEAN, edited by P.J.P.Whitehead *et al.*, 1984 and 1986, published by UNESCO. Fish names follow the same book, with additional information incorporated from the FAO SPECIES IDENTIFICATION SHEETS FOR FISHERY PURPOSES, 1981, edited by W.Fischer *et al.*, or the two Spanish books already mentioned. In a few cases, where there is more than one English name for a fish, we have used only the one listed by Whitehead. Where we give more than one Spanish name, the first is the official Spanish name and others are local names; sometimes the same local name is used for more than one species of fish. The sizes given for each species are maximum sizes recorded, so it can be assumed that most fish you see will be considerably smaller. We are indebted to Alberto Brito for local information on fish.

CARTILAGINOUS FISHES

Sharks and dogfish

Sharks and dogfish are predatory and depend mainly on smell for finding their food. Some species live on the floor of the continental shelf whilst others are entirely pelagic; many are confined to deep waters.

Family **LAMNIDAE - mackerel sharks**
Large fast swimming sharks with greyish-blue to black backs and white bellies. Can be dangerous.
Great White Shark, *Carcharodon carcharias*, ***Jaquetón Blanco, Jaquetón.***
Short Fin Mako, *Isurus oxyrinchus*, ***Marrajo Dientuso, Marrajo.***

Family **CARCHARHINIDAE - requiem sharks**
Small to medium sharks, frequently occurring close to the shore. Their backs vary in colour but they are always pale below. Dangerous to bathers.
Blue Shark, *Prionace glauca*, ***Tiburón Azul, Sarda, Quella.***
4.0m (13ft). This large and very elegant shark is an oceanic species that lives in the surface waters (down to about 200m), but it is also fairly frequent in coastal waters - especially when young. It is usually seen singly or in small groups, and sometimes accompanies boats; it cruises slowly, but is capable of bursts of speed. It is blue above and white below. It is a dangerous fish but apparently easy to catch; it is not a particularly well esteemed food.
Blacktip Shark, *Carcharhinus limbatus*, ***Tiburón Macuira, Jaqueta.***
Dusky Shark, *Carcharhinus obscurus*, ***Tiburón Arenero, Jaqueta.***
Over 2.5m (8.2ft). A semi-pelagic species which is frequently seen in coastal waters. It is grey-bronze above and white below. It is eaten.

Family **TRIAKIDAE - smooth dogfishes, houndsharks, smoothhounds**
Smoothhound, *Mustelus mustelus*, ***Musola, Cazón.***

Family **SPHYRNIDAE - hammerhead sharks**
Smooth Hammerhead, *Sphyrna zygaena*, ***Pez Martillo, Cornuda.***
4.0m (13ft). This species is often seen in coastal waters in the summer, usually arriving in June; it normally swims singly or in pairs and large individuals are very dangerous. It has a high dorsal fin which sticks out of the water as it swims near the surface, and a strange hammer-shaped head; it is deep olive to grey above and pale below. It is eaten.

Family **ALOPIIDAE - thresher sharks**
Thresher Shark, *Alopias vulpinus*, ***Zorro, Pejerrabo.***

Family **SCYLIORHINIDAE - catsharks**
Blackmouth Catshark, *Galeus melastomus*, ***Pintarroja, Bocanegra.***

Family **SQUALIDAE - dogfish sharks**
Gulper Shark, *Centrophorus granulosus*, **Quelvacho, Quelme.**
Velvet Belly, *Etompterus spinax*, **Negrito.**

Family **SQUATINIDAE - angel sharks**
Angelshark, *Squatina squatina*, **Angelote.**

Skates and rays

Several species of skates and rays are found in Canary waters and all live on the floor of the continental shelf, although a few also occur at greater depth on the continental slope. Like the sharks, they are predatory fish but they are dorsally compressed (flat) fish and typically prefer a soft sea bed where they spend some of their time partly covered with sand or mud. They are generally dark above and pale below. Most species are eaten.

Family **PRISTIDAE - sawfishes**
Smalltooth Sawfish, *Pristis pectinata*, **Pejepeine, Pez Sierra.**
Common Sawfish, *Pristis pristis*, **Pez Sierra Común.**

Family **RHINOBATIDAE - guitarfishes**
Blackchin Guitarfish, *Rhinobatos cemiculus*, **Guitarra Barbanegra, Guitarra.**
Common Guitarfish, *Rhinobatos rhinobatos*, **Guitarra Común, Guitarra.**

Family **TORPEDINIDAE - electric rays**
These fish have powerful electric organs on the head, capable of giving a shock of up to 250 volts.
Marbled Electric Ray, *Torpedo marmorata*, **Tembladera.**
> This fish occurs singly or in groups; it varies in colour but is usually mottled above.

Family **RAJIDAE**
Thornback Ray, *Raja clavata*, **Raya de Clavos, Raya.**

Family **DASYATIDAE - stingrays and whiprays**
Common Stingray, *Dasyatis pastinaca*, **Pastinaca Común, Chucho.**
> 2.5m (8.2ft). This species is found in shallow coastal waters on soft bottoms. It has a long tail with a serrated poisonous spine (whose effect is said to be excruciatingly painful) and is greyish, olive or brown above and whitish below. It is eaten.

Family **GYMNURIDAE - butterfly rays**
Spiny Butterfly Ray, *Gymnura altavela*, **Raya mariposa Espinuda, Mariposa.**

Family **MYLIOBATIDAE - eagle rays**
Common Eagle Ray, *Myliobatis aquila, **Aguila Marina, Pez Aguila, Ratón**.*
    1.5m (4.9ft). This species can sometimes be seen swimming in regular formations near the surface. It comes close to the shore for breeding in summer. It is dusky bronze above and has a long tail with a poisonous spine. It is eaten.
Bull Ray, *Pteromylaeus bovinus, **Chucho Vaca, Ratón, Obispo**.*

BONY FISHES

Family **CLUPEIDAE - herrings**
These are small to medium-sized laterally compressed pelagic shoaling fish; many feed on plankton, which they strain out of the water as it passes through a series of rakers on their gills. They are typically migratory, and the species form shoals in Canary waters and come close to the shore (mainly near the eastern islands) in the summer for spawning. They form the basis of important commercial fisheries. They are usually dark blue or blue-green on the back and have silvery flanks.
European Pilchard, *Sardina pilchardus, **Sarda Europea, Sardina**.*
Round Sardinella, *Sardinella aurita, **Alacha, Sardina Arencada**.*
Madeiran Sardinella, *Sardinella maderensis, **Machuelo**.*

Family **ENGRAULIDAE - anchovies**
European Anchovy, *Engraulis encrasicolus, **Anchoa Europea, Boquerón**.*

Family **AULOPIDAE - flagfins**
Royal Flagfin, *Aulopus filamentosus, **Lagarto Real, Lagarto de Hondura**.*

Family **SYNODONTIDAE - lizardfishes**
Solitary fish living in sandy, muddy or gravelly places. They stay very still on the sea floor, either propped up on their fins or partly buried in the sand, looking very reptile-like. They are metallic grey to brown on the back, with pale undersides. When unsuspecting prey comes close they rush out and usually swallow it in one gulp; sometimes they propel themselves right out of the water.
Red Lizardfish, *Synodus synodus, **Lagarto Diamante, Lagarto**.*
    0.4m (1.3ft). Grey with 4 large reddish blotches along the sides.
Atlantic Lizardfish, *Synodus saurus, **Pez de San Francisco, Lagarto**.*
    0.4m (1.3ft). Yellowish, reddish or brownish with 8 blotches along the sides. Usually in shallower water than the previous species.

Family - **ANGUILLIDAE - freshwater eels**
European Eel, *Anguilla anguilla, **Anguila**.*
    See FRESHWATER FISH.

Family - **MURAENIDAE - moray eels**

These are long and almost cylindrical fish without pelvic fins. Several species live in the area, but mainly around the eastern islands. Most of them are found only in fairly deep water. They spend the daytime hidden in cracks between rocks, in small caves or buried in the sand; usually only the head is visible. They are more active by night, when they come out to feed. These predatory fish can open their mouths incredibly wide, so that they can eat animals or parts of animals wider than their own diameter. They can be very dangerous and should never be provoked. They are mostly brown, or patterned brown. They are eaten.

Brown Moray, *Lycodontis unicolor*, ***Morena Lucia, Morena Morruda, Murión.***

0.95m (3.1ft). Medium to dark brown all over. Lives in shallow water on rocky, sandy or gravel bottom.

Mediterranean Moray, *Muraena helena*, ***Morena Mediterranea, Morena Pintada.***

Dotted Moray, *Muraena augusti*, ***Morena Augusta, Morena Negra.***

Fangtooth Moray, *Enchelycore anatina*, ***Morena Isleña, Bogavante, Abracante.***

Sharktooth Moray, *Gymnothorax maderensis*, ***Morena de Madeira, Murión de Hondura, Papudo.***

Family **CONGRIDAE - conger eels**

Elongate fish with cylindrical body and well developed eyes.

European Conger, *Conger conger*, ***Congrio Común, Congrio.***

2m (6.6ft), and can weigh as much as 30kg (66lb). Similar in shape and habits to the moray eels; greyish with white blotches It is dangerous and should never be approached casually.

Family **OPHICHTHIDAE - snake eels**

Leopard Eel, *Myrichthys pardalis*, ***Tieso Leopardo, Culebra, Carmelita.***

Family **BELONIDAE - needlefishes**

Long thin carnivorous fishes with long thin forceps-like mouths with many fine sharp teeth; they feed near the surface and are notable jumpers but do not actually "fly" like the flying fish.

Garfish, *Belone belone gracilis*, ***Agujón, Aguja.***

0.9m (3ft). A migratory shoaling pelagic species which arrives in coastal waters in summer. It is quite good to eat but the flesh is sometimes slightly green and the bones almost phosphorescent when cooked; this is said to discourage some people.

Family **SCOMBERESOCIDAE - sauries**

Surface-feeding fish; similar to the needlefishes but with the upper jaw much shorter than the lower. They feed on plankton animals.

Atlantic Saury, *Scomberesox saurus*, ***Paparda del Atlántico, Agujón, Paparda.***
> 0.46m (1.5ft). A pelagic shoaling species which inhabits surface waters, especially in inlets or coves; it frequently skips over the surface rather like a flyingfish. Olive to dark green or light brown; juveniles bluish. Is eaten.

Dwarf Saury, *Nanichthys simulans*, ***Paparda Enana, Paparda.***

Family **EXOCOETIDAE - flyingfishes**
The pectoral fins of these fish are expanded into thin membranes, and using these they glide through the air, having gained momentum by swimming up very fast through the water. This habit of "flying" out of the water when chased by predators, or when they confuse the vibration of a boat with a predator, means that they are more often seen than most other oceanic fish; they are frequently attracted by boat lights at night.

Tropical Two-wing Flying Fish, *Exocoetus volitans*, ***Pez Volador.***
> 0.3m (1ft). This pelagic shoaling fish is not normally seen close to the shore; it is, however, normally present all the year round further out to sea, especially in summer and autumn when the water is warmest, and can often be seen from boats. It is dark iridescent blue or green above and pale below.

Oceanic Two-wing Flyingfish, *Exocoetus obtusirostris*, ***Volador Ñato.***
Bennett's Flyingfish, *Cypselurus pinnatibarbatus*, ***Volador de Bennett.***

Family **HEMIRAMPHIDAE - halfbeaks**
Balao Halfbeak, *Hemiramphus balao*, ***Agujeta Balajú, Aguja.***

Family **MACRORAMPHOSIDAE - snipefishes**
Longspine Snipefish, *Macroramphosus scolopax*, ***Trompetero.***

Family **AULOSTOMATIDAE - trumpetfishes**
Fish in this group are elongate and all have a tube-shaped snout with a small mouth at the end, through which they suck small fish and crustaceans by sudden expansion of the mouth cavity.

Trumpetfish, *Aulostomus strigosus*, ***Escopeta, Pez Trompeta.***
> 0.65m (2.1ft). Found in rocky places near the shore.

Family **SYNGNATHIDAE - pipefishes and seahorses**
These have several features in common with the Aulostomatidae but have their bodies surrounded by bony rings.

a seahorse, *Hippocampus ramulosus*, -- ***Caballo Marino, Cabillito de Mar.***
a pipefish, *Syngnathus acus*, —, ***Peje Pipa.***

Family **MACROURIDAE - grenadiers**
Blackspot Grenadier, *Coelorhynchus coelorhyncus*, **Granadero Aco-razado, Ratón.**
Softhead Grenadier, *Malacocephalus laevis*, **Abámbolo de Bajura.**
Smooth Grenadier, *Nezumia aequalis*, **Granadero Lisa.**
Roughsnout Grenadier, *Trachyrincus scabrus (T.trachyrincus)*, **Abámbolo de Cantil.**

Family **GADIDAE - cods**
Greater Forkbeard, *Phycis blennoides*, **Brótola de Fango, Brota.**
Forkbeard, *Phycis phycis*, **Brótola de Roca, Brota.**

Family **MORIDAE - morid cods**
Common Mora, *Mora mora*, **Mollera Moranella, Hediondo.**

Family **LAMPRIDIDAE - opah**
Opah, Moonfish, *Lampris guttatus*, **Opa, Luna Real.**

Family **POLYMIXIDAE - beardfishes**
Beardfish, *Polymixia nobilis*, **Salmón de lo Alto.**

Family **BERYCIDAE - alfonsinos**
Slender Alfonsino, *Beryx splendens*, **Alfonsino Besugo, Besugo Ameri-cano, Alfonsino, Fula Colorada.**
Alfonsino, *Beryx decadactylus*, **Alfonsino Palometón, Palometa Roja, Afonsino, Fula Colorada.**

Family **TRACHICHTHYIDAE - slimeheads**
Darwin's Slimehead, *Gephyroberyx darwini*, **Reloj de Darwin.**
Mediterranean Slimehead, *Hoplostethus mediterraneus*, **Reloj Medit-erráneo.**

Family **ZEIDAE - dories**
John Dory, *Zeus faber*, **Pez de San Pedro, Gallo Barbero.**

Family **SERRANIDAE - sea basses**
This family includes many of the most important food fish of tropical and temperate seas. They vary in length from a few cm to about 1.8m (6ft); most are carnivorous and many species sit on the sea floor waiting for their prey, whilst others swim in shoals.
Dusky Grouper, *Epinephelus guaza*, **Mero.**
    1.5m (4.9ft). A solitary fish that stays in its rocky lair on the sea bed during the day and comes out to hunt octopus and cuttlefish at night. It has a greyish or reddish-brown back, scattered with white spots; the belly is pale yellow. It is a popular fish with divers, since it is inquisitive and "friendly". A particularly good food fish.

Comb Grouper, *Mycteroperca rubra*, ***Gitano, Abade, Abae.***
>0.9m (3ft). This fish is rather similar to the previous species, and sometimes swims in shoals of up to several hundred. It is a particularly good food fish.

Blacktail Comber, *Serranus atricauda*, ***Serrano Imperial, Cabrilla.***
>0.35m (1.1ft). An attractive fish living in rocky areas. It has 4 or 5 squarish dark blotches along its sides. A good food fish.

Comber, *Serranus cabrilla*, ***Cabrilla, Cabrilla Reina, Cabrilla Rubia.***
>0.28m (0.9ft). Similar to above, but not so common. It is a very good food fish.

Painted Comber, *Serranus scriba*, ***Serrano Escribano, Serrano, Vaquita.***

Wreckfish, *Polyprion americanus*, ***Cherna, Cherne.***

Swallowtail Seaperch, *Anthias anthias*, ***Tres Colas.***

Family **MORONIDAE - temperate basses**
European Seabass, *Dicentrarchus labrax*, ***Lubina, Róbalo.***
Spotted Seabass, *Dicentrarchus punctatus*, ***Baila.***

Family **PRIACANTHIDAE - bigeyes**
Bright red deep-bodied nocturnal fish with large red eyes; they are carnivouous, feeding mainly at night, and live on the sea-bed in shallow water.

Glasseye, *Priacanthus cruentatus*, ***Catalufa de Roca, Catalufa.***
>0.3m (1ft). A solitary fish living on rocky bottoms in shallow water. Very good to eat.

Family **APOGONIDAE - cardinalfishes**
Small tropical and subtropical brightly coloured fish with fairly deep compressed bodies and large eyes. They are bottom-dwelling, largely nocturnal and mostly found in shallow water.

Cardinal Fish, *Apogon imberbis*, ***Salmonete Real, Funfurriña, Alfonsito.***
>0.1m (0.3ft). This bright red fish is one of the commonest species in many sea caves. It is an excellent food fish.

Family **POMATOMIDAE - bluefish**
Fast swimming schooling fish; they are reputed to be extremely voracious and often kill more fish than they can eat.

Bluefish, *Pomatomus saltator*, ***Anchova de Banco, Pejerrey, Anjova.***
>1.1m (3.6ft). This fish is found swimming in groups around offshore rocks and banks. It is a greenish or bluish-grey fish with silvery sides and belly, and yellow on its fins. Good food fish with dark meat.

Family **CARANGIDAE - jacks and pompanos etc**

This family includes fish of very different shapes, but they are usually dark above and pale below; most are schooling species.

Greater Amberjack, *Seriola dumerili*, *Pez de Limón, Medregal.*

1.7m (5.6ft). A beautiful bluish-grey fish with silvery vertical bands. Swims in pairs or groups; its persistent attention to divers and underwater photographers can become a nuisance. Good food fish with dark meat.

Guelly Jack, *Pseudocaranx dentex*, *Jurel Dentón, Jurel.*

0.8m (2.6ft). This is is a coastal shoaling species which feeds both in open water and on the seabed. It is a beautiful strong-swimming and very active fish; deep-bodied and elongate, pale greenish above and silvery below. A good food fish with dark meat.

Derbio, *Trachinotus ovatus*, *Palometa Blanca, Palometa.*

0.45m (1.5ft). A silvery grey fast-swimming shoaling fish which is particularly common around rocks in coastal waters in summer and autumn. Is eaten.

Blue Jack Mackerel, *Trachurus picturatus*, *Jurel de Altura, Chicharro.*

0.45m (1.5ft). A strong-swimming fish that comes inshore for spawning, and swims closer to the surface at night. It has a dark head and paler whitish to silvery back. Is eaten.

Atlantic Horse Mackerel, *Trachurus trachurus*, *Jurel, Chicharro.*

Vadigo, *Campogramma glaycos*, *Lirio.*

Leerfish, *Lichia amia*, *Palometón.*

Pilotfish, *Naucrates ductor*, *Pez Piloto.*

Family **CORYPHAENIDAE - dolphinfish**

Fast-swimming carnivorous oceanic fish with elongate and compressed bodies and large mouths. The adult males have a bony crest on the front of the head. They are popular with sport fishermen; swim singly or in shoals.

Common Dolphinfish, *Coryphaena hippurus*, *Llampuga, Dorado.*

2.0m (6.6ft). This fish comes close to the shore in summer to breed. It swims close to the surface and sometimes can be seen under boats or other floating objects. Its beautiful blue colour fades very quickly after death.

Pompano Dolphinfish, *Coryphaena equiselis*, *Dorado.*

Family **BRAMIDAE - pomfrets**

Atlantic Pomfret, *Brama brama*, *Japuta, Pez Tostón.*

Big Scale Pomfret, *Taractichthys longipinnis*, *Cangullo, Pez Tostón Volador.*

Family **POMADASYIDAE (HAEMULIDAE) - grunts**

These are oblong perch-like fish, vary variable in colour. They make a grunting noise by grinding their teeth, the sound being amplified by the swim bladder.

Bastard Grunt, *Pomadasys incisus*, **R**onco Mestizo, Roncador, Ronco, Tronelero.

0.3m (1ft). Large shoals swim in coastal waters over rocky bottoms. Background colour is silvery-grey, sometimes with large blotches but never with stripes or spots. A good food fish.

African Striped Grunt, *Parapristipoma octolineatum*, **Burro Listado, Boca de Oro, Burrito.**

0.4m (1.3ft). This shoaling fish is found in shallow waters over rocky and sandy bottoms. It is violet-brown with 4 longitudinal blue stripes along back and sides. A good food fish.

Family **SCIAENIDAE - drums**

These are medium sized carnivouous fish that live in fairly shallow water. They make a variety of croaking sounds, using their swim bladders as resonators.

Meagre, *Argyrosomus regius*, **Corvina.**

An excellent food fish, but only rarely found around Tenerife.

Brown Meagre, *Sciaena umbra*, **Corvallo, Verrugato.**

Canary Drum, *Umbrina canariensis*, **Verrugato de Canarias, Verrugato.**

Shi Drum, *Umbrina cirrosa*, **Verrugato Común, Verrugato.**

Fusca Drum, *Umbrina ronchus*, **Verrugato Fusco, Verrugato.**

Family **MULLIDAE - goatfishes or surmullets**

These somewhat elongate fish have barbs under the chin that are tactile organs for detecting their invertebrate food, and are constantly stirring up the sand as they search.

Striped Red Mullet, *Mullus surmuletus*, **Salmonete de Roca, Salmón.**

0.4m (1.3ft). Pinkish fish that live in small groups in muddy and sandy zones and feed on molluscs, crustaceans and worms. The species is seen close to the shore in spring and again in October. A good food fish.

Family **SPARIDAE - porgies etc**

Mostly deep-bodied carnivorous fish, usually with sharp teeth. Many are popular angling fish.

Common Dentex, *Dentex dentex*, **Dentón Común, Sama de Ley, Sama Dorada.**

1.0m (3.3ft). This is a coastal species, found over hard bottoms. Like all fish in this genus, it changes its shape and colour as it gets older; it has greyish young, which become pink as they mature and bluish-grey with dark spots when they are old. A good food fish.

Pink Dentex, *Dentex gibbosus,* **Sama de Pluma, Pargo Macho, Serruda.**
1.0m (3.3ft). Found in coastal areas over rocky bottoms. A reddish fish with bluish-silvery reflections. Large individuals are now rare because of overfishing. A good food fish.

Large-eyed Dentex, *Dentex macrophthalmus,* **Cachucho, Antonito.**

Redbanded Seabream, *Sparus auriga,* **Pargo Sámola, Sama Roquera.**
0.6m (2.0ft). Young of this species live near the coast over hard bottoms, while adults live further out. It is silvery-pink with 4 or 5 vertical bands. A good food fish.

Bluespotted Seabream, *Sparus caeruleostictus,* **Zapata, Catalineja, Hurta, Roquera.**
0.7m (2.3ft). Swims singly or in small groups over hard bottoms; young are found closer to shore. Varies in coloration, but often silver-pink with bluish-black spots. It feeds partly on bivalves which it crushes with its jaws. It is easy to catch and because of this is rarer on calm shores. A good food fish.

Common Seabream, *Sparus pagrus pagrus,* **Pargo, Bocinegro, Pallete (young ones only).**
0.75m (2.5ft). A silvery-pink fish similar in appearance and habits to the previous species. A very popular food fish.

Gilthead Seabream, *Sparus aurata,* **Pargo Dorado, Dorada.**

Common Pandora, *Pagellus erythrinus,* **Breca.**
0.6m (2ft). A pink fish with small blue spots that lives on various types of seabed along the coast. The numbers are declining because of overfishing, and most individuals that are caught are fairly small. A good food fish.

Axillary Seabream, *Pagellus acarne,* **Aligote, Besugo.**

Blackspot Seabream, *Pagellus bogaraveo,* **Goraz.**

Saddlebream, *Oblada melanura,* **Oblada, Galana.**
0.3m (1ft). This species is common around coastal rocks in shallow water, but elusive; it swims in shoals of up to 100. It is silvery blue-grey with a large black spot at the base of the tail. Is eaten.

Annular Seabream, *Diplodus annularis,* **Raspallón, Anillado, Mojarra, Rufiana, Amarillo, Mugarra.**
0.2m (0.7ft). This species lives entirely in beds of green algae on rocks close to the shore. It is a greyish fish tinged with yellow and with yellow fins. Is eaten.

Common Two-banded Seabream, *Diplodus vulgaris*, **Sargo, Mojarra, Seifia, Saifia, Saifio.**

0.45m (1.5ft). A very common solitary fish, living among rocks in fairly shallow water over rocky or sandy bottoms. It is grey, brownish or greenish with a large dark "saddle" behind its head. A good food fish.

White Seabream, *Diplodus sargus cadenati*, **Sargo Marroqui, Sargo Blanco, Sargo.**

0.45m (1.5ft). One of the commonest fish around the shores, swiming over rocky bottoms in schools. It has dark vertical bands on its silvery-grey body. An excellent food fish.

Zebra Seabream, *Diplodus cervinus cervinus*, **Sargo Soldado, Sargo Breado.**

0.55m (1.8ft). This species lives among rocks and swims in small mixed-size groups, which are thought to be family groupings. Its body has very strong zebra-like bands. A good food fish.

Sharpsnout Seabream, *Diplodus puntazzo*, **Sargo Picudo, Morruda.**

0.6m (2ft). This fish can be seen in small groups in fairly shallow water over rocky bottoms. The young are often found in rock pools and the adults in the surf zone. It is silvery-grey with alternate dark and light bars on its sides. A good food fish.

Striped Seabream, *Lithognathus mormyrus*, **Herrera.**

0.55m (1.8ft). This species swims in groups of mixed sizes (possibly family groups) for much the year, but forms big shoals in summer. It prefers sandy or muddy bottoms and feeds on molluscs and crustaceans and produces clouds of sand as it hunts for them. It is marked with vertical bands. A good food fish.

Salema, *Sarpa salpa*, **Salema.**

0.45m (1.5ft). This species feeds on algae close to the shore, either over rocky or sandy bottoms. Individuals sometimes get trapped in rock pools at low tide. They are greyish-blue with thin orange-golden longitudinal stripes. A good food fish.

Bogue, *Boops boops*, **Boga.**

0.36m (1.2ft). This fish form huge shoals close to the shore and feeds primarily on plankton. It is not deep bodied like the others in this family. Its back is bluish or greenish, and the sides have silvery or golden reflections. Is eaten.

Black Seabream, *Spondyliosoma cantharus*, **Chopa Negra, Chopa.**

0.6m (2ft). A pelagic fish that comes close to the coast to spawn. It sometimes forms shoals of thousands of individuals. The young are found in shallow water among seaweed. It is silver-grey with bluish, greenish or pinkish reflections; golden-yellow longitudinal lines run along its sides. A good food fish.

Family **CENTRACANTHIDAE - picarels**
Curled Picarel, *Centracanthus cirrus*, *Jerret Imperial*, *Madre de la Boga*.

Family **POMACENTRIDAE - damselfish**
Primarily a tropical family. Deep bodied, laterally compressed and fairly small fish, often with considerable colour variation within a species.
Bandtail Chromis, *Chromis limbatus*, *Castāneta Rabo-cinta*, *Fula Blanca*.
　0.12m (0.4ft). A shoaling species, living in rocky and sandy areas. It becomes territorial during the breeding season. It is golden-brown with transparent yellow pectoral fins. Is eaten.
—, *Abudefduf luridus*, **Castāneta Negra, Fula Negra.**
　0.12m (0.4ft). Young of this species are often found in rock pools, and are brilliant cobalt blue; adults are similar to the previous species. Is eaten.

Family **LABRIDAE - wrasses**
A large and varied family of carnivorous fishes; many are brightly coloured.
Barred Hogfish, *Bodianus scrofa*, **Vieja, Pejeperro.**
　0.43m (1.4ft). A longish fish with pointed head. It swims solitarily or in pairs, searching among rocks for its prey of small fish, octopus, squid and crustaceans. It has formidable sharp teeth. The males are red and the females red above and yellow below. The males are the best to eat. Although the official Spanish name for this fish is *Vieja*, in the Canary Islands that name is used for the Parrotfish.
—, *Centrolabrus trutta*, **Romero.**
　This fish lives in shallow water close to the shore. It varies in colour but some are a deep green.
Rainbow Wrasse, *Coris julis*, **Julia, Señorita, Doncella.**
　0.25m (0.8ft). A beautiful longish fish, variable in colour, that lives around coastal rocks in small groups.
Ornate Wrasse, *Thalassoma pavo*, **Fredi, Pez Verde, Pejeverde.**
　0.2m (0.7ft). A very beautiful fish that lives in small groups close to the shore in rocky areas, especially where there are algae. Small individuals are common in rock pools. Its colours and pattern differ according to age, time of year and water depth, but green predominates. Is eaten.
Cleaver Wrass, *Xyrichthys novacula*, **Doncella Cuchilla, Raó, Raór, Pejepeine, Vaqueta.**
　0.25m (0.8ft). An inshore fish inhabiting sandy and stony areas. It buries itself in times of danger. Is eaten.

Ballan Wrass, *Labrus bergylta*, **Vaqueta, Romero Capitán.**
Scale-eyed Wrass, *Acantholabrus palloni*, **Tac Rocas.**

Family **SCARIDAE - parrotfishes**
These fishes have different colour phases, varying from very dark to reddish; the sexes are also different. They are creatures of habit and return to regular sleeping places for the night. They graze on the rocks, biting off bits which they then crush with plate-like teeth at the back of their throats.
Parrotfish, *Sparisoma cretense*, **Loro Viejo, Vieja.**
    0.5m (1.6ft). An abundant inshore fish, which inhabits rocky places and thick patches of alga; it lives in small mixed age groups. It is very variable in colour. A very popular food fish; it is locally known as **Vieja**, which is also the Spanish name for the **Barred Hogfish.**

Family **TRACHINIDAE - weeverfishes**
A small family of elongate fish which live in shallow water and often bury themselves in sand or mud. They are very dangerous fish with poisonous spines, containing a powerful blood and nerve toxin; it takes several months to recover from the effects.
Greater Weever, *Trachinus draco*, **Escorpión, Araña.**
    0.45m (1.5ft). The back of this burrowing fish is greenish and it has fine oblique bands on the sides of the body. A good food fish needing careful preparation.
Streaked Weever, *Trachinus radiatus*, **Víbora, Araña.**

Family **GEMPYLIDAE - snake mackerels**
Escolar, *Lepidocybium flavobrunneum*, **Escolar Negro, Escolar Chino.**
Promethian Escolar (Rabbitfish), *Promethichthys prometheus*, **Escolar Prometeo, Conejo.**
Oilfish, *Ruvettus pretiosus*, **Escolar Clavo, Escolar.**
    This is not a popular food fish due to the fact that its flesh has a strong purgative effect.

Family **TRICHIURIDAE - cutlassfishes**
Black Scabbardfish, *Aphanopus carbo*, **Sable Negro.**
Silver Scabbardfish, *Lepidopus caudatus*, **Pez Cinto, Sable.**
Largehead Hairtail, *Trichiurus lepturus*, **Pez Sable.**

Family **SCOMBRIDAE - tunas and mackerel**
Fast-swimming carnivorous fish with streamlined bodies which taper to a narrow "neck" at the base of the tail fin. Fisheries in many parts of the world depend upon this group.
Atlantic Bluefin Tuna, *Thunnus thynnus*, **Atún Rojo, Patudo.**
    3.1m (10.2ft). A migratory oceanic fish with dark blue or black back and

sides, and silvery-white belly. This very large fish is popular with sport fishermen because of the good fight it puts up when hooked. Is eaten.

Bigeye Tuna, *Thunnus obesus*, **Patudo, Tuna.**
2.4m (7.9ft). A pelagic oceanic species. It has metallic dark blue back and fins tinged with yellow. It is eaten.

Yellowfin Tuna, *Thunnus albacares*, **Rabil.**

Albacore, *Thunnus alalunga*, **Atún Blanco, Albacora.**

Skipjack Tuna, *Katsuwonus pelamis*, **Listado, Bonito.**
1.03m (3.4ft). A migratory oceanic species. It has a dark purplish-blue back, silvery lower sides and belly and 4-6 longitudinal dark bands. It is a popular food fish.

Chub Mackerel, *Scomber japonicus*, **Estornino, Caballa.**
0.52m (1.7ft). An elongate pelagic tuna which comes to the shore in the spring and summer for breeding. Its back is steel blue crossed with faint wavy lines and the lower sides and belly are silvery-yellow with dusky rounded blotches.

Atlantic Mackerel, *Scomber scombrus*, **Caballa del Atlántico.**

West African Spanish Mackerel, *Scomberomorus tritor*, **Carite Pintado, Carita.**

Plain Bonito, *Orcynopsis unicolor*, **Tasarte.**

Atlantic Bonito, *Sarda sarda*, **Bonito Atlántico, Sierra.**

Family **ISTIOPHORIDAE - sailfishes**
Blue Marlin, *Makaira nigricans*, **Marlin Azúl, Ajuija.**

Family **XIPHIIDAE - swordfishes**
Swordfish, *Xiphias gladius*, **Pez Espada.**

Family **GOBIIDAE - gobies**
A very large family of small to medium-sized bottom-living fish - locally called ***Cabosos***. They have rounded heads and their eyes are usually close together: their pelvic fins are modified to form a sucker.

—, *Mauligobius maderensis*, —, ***Caboso***
A small fish, endemic to the Canaries, Madeira and the Salvages. It lives in rock pools, where it hides in cracks and below stones. It is able to jump out of a pool and jerk its way across the rocks for several metres; it can survive for a while out of water.

Rock Goby, *Gobius paganellus*, **Bobi, Caboso.**
Similar to above but also found lower down on the shore.

Black Goby, *Gobius niger*, **Chaparrudo, Caboso Negro.**
Although the larvae of this fish can be found in rock pools, the adults live further from the shore.

Family **BLENNIIDAE - combtooth blennies**

Small elongate carnivorous or omnivorous bottom-dwelling fish - locally called *barrigudas*. They are generally found in shallow water.

—, *Blennius parvicornis*, —, **Barriguda.**
   Rarely more than 0.1m (0.3ft). A small blunt-headed fish, with small filaments on its head; commonly found in rock pools. It is able to jump out of a pool and jerk its way across the rocks for several metres; it can survive for a while out of water, its skin being protected against dessication.

— *Blennius trigloides*, —, **Barriguda.**
   Similar to above but found lower on the shore.

—, *Blennius cristatus*, —, **Barriguda.**
   Rock pools. Similar to above.

—, *Coryphoblennium galerita*, —, **Barriguda.**
   Similar to above.

— *Ophioblennius atlanticus atlanticus*, —, **Barriguda Negra, Bariguda Mora.**
   Similar to above but found mainly below the low tide mark.

Family **CENTROLOPHIDAE - medusafishes**
Imperial Blackfish, *Schedophilus ovalis*, **Rufo Imperial, Pampano.**

Family **SPHYRAENIDAE - barracudas**
Inquisitive predatory fish that hunt by sight rather than smell, and investigate anything brightly coloured or moving. They are elongate, with a large head and pointed snout.

European Barracuda, *Sphyraena sphyraena*, **Espetón, Bicuda.**
   1.65m (5.4ft). This is a pelagic coastal species that forms large shoals. Its back is bluish-grey and it is silvery white below; there are short vertical bands on its sides. Barracudas sometimes attack divers. Is eaten.

Yellowmouth Barracuda, *Sphyraena viridensis*, **Espetón Boca Amarilla, Bicuda.**

Family **MUGILLIDAE - mullets**
Torpedo-shaped, shallow-water schooling fish, living in muddy and sandy areas. They have small mouths and are primarily vegetarian and grub around for food.

Flathead Grey Mullet, *Mugil cephalus*, **Pardete, Lisa, Lebrancho.**
   1.20m (3.9ft). Back greyish blue, sides and belly silvery-whitish, sometimes with golden reflections. Is eaten.

Golden Grey Mullet, *Liza aurata*, **Galupe, Lisa Amarilla.**
   0.45m (1.5ft). Inhabits shallow coastal waters. It has a greyish-brown back and its sides and belly are silvery. Is eaten.

Thicklip Grey Mullet, *Chelon labrosus*, **Lisa Negra, Lisa, Cabezote.**

Family **ATHERINIDAE - silversides**
Small greenish or bluish elongate fish with a conspicuous silvery stripe along the side.

Sand Smelt, *Atherina presbyter*, *Pejerrey, Guelde, Longorón.*

> 0.14m (0.5ft). An entirely coastal species, found especially in coves and around underwater cliffs. It has been overfished and most individuals now caught are small. A common food fish.

Family **SCORPAENIDAE - scorpion fishes**
These are all bottom-living carnivorous fish, found in rocky, gravelly or sandy places; only two species are often found close to the shore. The family includes a species which has the distinction of being the most poisonous fish in the world. They all have a dorsally compressed body and poisonous spines. Most are live bearers.

Madeira Rockfish, *Scorpaena maderensis*, *Poyo, Rascacio de Madeira, Rascacio, Rocas.*

> 0.2m (0.7ft). Stays still at cave entrances; can change colour. It is eaten.

Black Scorpion Fish, *Scorpaena porcus*, *Escorpora, Rascacio Negro, Rascacio, Rocas.*

> 0.23m (0.8ft). Similar to previous species.

Yellow-orange Scorpion Fish, *Pontius kuhlii*, *Rascacio Amarillo, Obispo, Volón.*

Red Scorpion Fish, *Scorpaena scrofa*, *Cabracho, Cabrachgo, Cantarero.*

Blackbelly Rosefish, *Helicolenus dactylopterus dactylopterus*, *Gallineta, Bocanegra.*

Family **TRINGLIDAE - unarmoured searobins**
Streaked Gurnard, *Triglophorus lastoviza*, *Rubio.*
Tub Gurnard, *Trigla lucerna*, *Bejel, Rubio.*

Family **DACTYLOPTERIDAE - flying gurnards**
Flying Gurnard, *Dactylopterus volitans*, *Pez Volador, Alón, Chicharra.*

Family **BOTHIDAE - lefteye flounders**
Flat, bottom-living predatory fish. They, in effect, lie on their sides, although both their eyes are on the upper side.
Wide-eyed Flounder, *Bothus podas maderensis*, *Podas, Tapaculo.*

Family **SOLEIDAE - soles**
Sand Sole, *Solea lascaris*, *Sortijo, Lenguado.*
Common Sole, *Solea vulgaris*, *Lenguado Común, Lenguado.*

Family **BALISTIDAE - triggerfishes**

A group of strange-looking deep-bodied carnivorous fish with small mouths and strong teeth; many are poisonous. They have thick tough skins. The triggerfishes can jam themselves into holes in the rock by locking a long dorsal spine into an upright position. Some make a grunting sound when removed from water.

Grey Triggerfish, *Balistes carolinensis*, **Pejepuerco Blanco, Pez Ballesta, Gallo.**
> 0.45m (1.8ft). This solitary fish is found in shallow water in sandy, stony and gravelly areas. Delicious to eat.

Family **MONACANTHIDAE - filefishes**

Bottom-living fish similar to the Balistidae, but with narrow bodies.

— *Stephanolepis hispidus*, —, **Gallo, Gallo Cochino.**
> 0.26m (0.9ft). Solitary, but often abundant in stony, gravelly and sandy areas and alga zones. Swims slowly without fear and in apparent confidence in its defence. Is eaten, but care needed with preparation.

Family **TETRAODONTIDAE - puffers**

Small to medium carnivorous fish with heavy blunt bodies. When disturbed, can swallow either water or air and swell up into a sphere. The flesh of many species contains a deadly poison.

— *Sphoeroides spengleri*, —, **Tamboríl.**
> ca.0.15m (ca.0.5ft). Along rocky shores and in beds of green algae.

—, *Sphoeroides cutaneus*, —, **Tamboríl de Hondura.**

— *Canthigaster rostrata*, —, **Gallinita.**
> Very common.

Prickly Puffer, *Ephippion guttifer*, —, **Tamboríl de Tierra.**

Smooth Puffer, *Lagocephalus laevigatus*, **Tamboríl, Tamboríl Mondeque.**

Family **GOBIESOCIDAE - clingfishes**

Small fish with broad heads and slender bodies. The pelvic fins are modified to form a sucking disc on the ventral surface so that they can cling firmly to rocks and seaweed. They live in shallow water mainly in the intertidal zone.

— *Lepadogaster lepadogaster lepadogaster*, —, **Chupasangre, Pégalo.**

— *Lepadogaster zebrina*, —, **Chupasangre, Pégalo.**
> Endemic to the Canary Islands and Maderia.

Connemarra Clingfish, *Lepadogaster candollei*, —, **Chupasangre.**

SELECTED LAND INVERTEBRATES

There are a very large number of land invertebrate animals on Tenerife and, apart from the dragonflies, butterflies and hawkmoths we make no attempt to give comprehensive lists. We do, however, mention all the major groups (and some small ones of special interest) and try to give an idea of their importance on the island. We also give some information on all the species that are mentioned in the text of Sections 1 and 2.

MOLLUSCA

TERRESTRIAL GASTROPODS - snails and slugs

Only a small number of stocks of snails seem to have colonized the Canaries naturally, but those that did so have tended to diversify. Two genera (*Hemicycla* and *Canariella*) are endemic to the Canaries and another three are found only in the Canaries and other Macaronesian islands. Some of these genera have given rise to a large number of distinct forms within the islands, although in many cases it is not clear which of them should be considered as separate species. In the case of the slugs, it seems likely that most species have been introduced by humans.

Hemicycla bidentalis (Family Helicidae).

> Member of a Canary endemic genus. This species is a dark brown and dull orange snail which may be found in the laurel forest.

Insulivitrina lamarcki (Phenacolimax lamarcki) (Family Vitrinidae).

> Endemic species belonging to a Macaronesian endemic genus, sometimes considered a subgenus of *Phenacolimax*. This slug-like animal is really a snail in which the shell is much reduced and partially internal; found in the laurel forest.

ANNELIDA

OLIGOCHAETA - earthworms

Tenerife is rich in earthworms, but a recent study shows that many of the species are of wide distribution and that the fauna includes tropical as well as northern species; a large proportion may well have been accidentally introduced by humans along with imported plants.

ARACHNIDA

SCORPIONES - scorpions

Scorpions are poor dispersers, and none have colonized the Canaries naturally; there is, however, a single introduced species.

Centruroides nigrescens (Family Buthidae).

This species is found only in the coastal area close to Santa Cruz de Tenerife, and was evidently brought in accidentally on a ship. Its sting contains a powerful poison, but it is not normally fatal to humans.

SCHIZOMIDA - schizomids

This obscure group is mentioned only because of the recent interesting discovery of a population of the species *Schizomus portoricensis* in two volcanic caves in Tenerife; only females are present, these being capable of asexual reproduction. This species was previously unknown on the eastern side of the Atlantic, except in an orchid house in Cambridge!

ARANEAE - spiders

This is by far the most important group of arachnids on the Canaries, with about 160 species recorded from Tenerife, in 26 families. The spiders include a number of species with wide distribution outside the Canaries, along with many endemic species. A technical book on the spiders of the Canaries and Madeira has recently been published (mainly in German) by Jörg Wunderlich (see APPENDIX, Books); he describes 98 species new to science from the Canaries, but there are certainly plenty more still to be found.

Dysdera spp. (Family Dysderidae).

This genus of ground-dwelling spiders, typically red-brown in colour, has undergone an evolutionary radiation on the Canaries, and new species are still being found. On Tenerife they occur in many habitats, including caves, but can most easily be found under rocks or logs, often with their large silky white egg sacs.

Latrodectus tredecimguttatus (L. mactans tredecimguttatus), **Black Widow Spider**, *Araña Negra, Mamona* (Family Theridiidae).

The species found in the Canaries also occurs in Madeira, the Mediterranean region and northern Africa, but its American relatives are now considered to belong to a different species. This poisonous black spider is uncommon on Tenerife, but has been found occasionally near Santa Cruz. It usually makes its web among loose rocks, and may be confused with the related and much commoner - but non-poisonous - *Steatoda* species. It is best to leave severely alone any dark coloured spider with a more or less spherical abdomen.

Argyrodes argyrodes (Conopistha argyrodes) (Family Theridiidae).

Not endemic; widespread in the Old World. A minute silvery spider that lives in the webs of *Cyrtophora citricola* (see below).

Lepthyphantes oromii (Family Linyphiidae).

An endemic blind cave spider, whitish and quite small, only recently discovered and described.

Cyrtophora citricola (Family Araneidae).

Not endemic; widespread in warm areas of the Old World. The females of this orb-weaver are large and black with white markings, but the males are much smaller. *Argyrodes* (see above) can sometimes be found in the webs. Bushes in many parts of the lower zone, including gardens, are often festooned with the webs of this species.

Agelena canariensis (Family Agelenidae).

Not endemic; occurs in the Canaries and Morroco. This funnel-weaver is a close relative of the long-legged dark brown spiders that frequently appear in houses in northern Europe, especially in autumn. Its webs are often conspicuous on banks and around the bases of trees in the forest zone.

Cheiracanthium pelasgicum (Family Clubionidae).

Not endemic; occurs in the Mediterranean region and in West Africa. A large yellowish spider that occurs in the leaf litter of the laurel forests. Its bite may be poisonous.

Olios canariensis (O.argelasius) (Family Heteropodidae).

Macaronesia endemic. This giant crab spider is one of the most impressive spiders present on the island. Its body is very much flattened and it can be found in crevices in the bark of very old specimens of **Canary Pine**.

Phlegra lucasi (Aelurillus restingae) (Family Salticidae).

Not endemic. This small, dark brown jumping spider can be found on lava flows and cinder fields in the high mountain zone: recorded from an altitude of 3300m (10800ft) above the *Refugio de Altavista.*

OPILIONES - harvestmen

The harvestmen of the Canaries are remarkable in that all three known species are endemic, and form two endemic genera, *Parascleropio* and *Bunochelis*; evidently the colonization of the islands by this group occurred a very long time ago. The genus *Parascleropio* is of special interest, in that its closest relatives are now found in the Balkans: presumably its ancestors also occurred in northwest Africa.

Bunochelis spinifera (Family Phalangiidae).

Canary endemic. Widespread on Tenerife, and can be found at the summit of Mt.Teide.

PSEUDOSCORPIONES - false scorpions

About 20 species of false scorpions occur in the Canaries, and Tenerife has the richest fauna. This includes two endemic genera, *Pseudorhaco-chelifer* and *Canarichelifer*, each with a single species.

CRUSTACEA - crustaceans

A number of species of crustaceans that live as plankton in freshwater have been found in the Canaries, most being North African species. The terrestrial forms are of more interest and the woodlice (Isopoda) have diversified within the archipelago, although there are some that also occur in the Mediterranean region. Tenerife has 21 species, four of which are endemic: one of these was discovered in Anaga only a few years ago.

ISOPODA - woodlice etc.

Haplopthalmus danicus (Family Trichoniscidae).

Not endemic. Found in caves but also occurs in leaf litter in the laurel forests of the Anaga peninsula.

AMPHIPODA

Orchestia chevreuxi (Family Talitridae).

Macaronesian endemic. This is a terrestrial species, occurring in leaf litter in moist places in the forests along the north of the island.

DIPLOPODA - millipedes

Although only about 20 species of millipedes have been found in the Canaries, a strikingly high proportion of these are endemic: Tenerife has several different species of the genus *Dolichoiulus*, which have evidently evolved on the island.

CHILOPODA - centipedes

There are only a few species of centipedes in the Canaries, and most of them are widely distributed outside the archipelago.

Lithobius pilicornis (Family Lithobiidae).

Not endemic; also occurs in western Europe, northern Africa, and other Macaronesian islands. Found in caves, but also occurs above ground.

Lithobius crassipes (Family Lithobiidae).

Not endemic; widespread in Europe, western Asia and northern Africa. This species is able to live even in barren lava flows, and we have found it within a few metres of the summit of Mt.Teide.

Scolopendra morsitans, **Escolopendra** (Family Scolopendridae).

Not endemic. This centipede is large and brightly coloured (grey-blue with yellow legs), and is one of the more impressive invertebrates on the island. It is common in the lower zone, and can also be found in the pine forest, usually under rocks or logs. It can inflict a painful bite if handled, but will run fast into cover if given a chance.

Scutigera coleoptrata (Family Scutigeridae).

Not endemic; it is also found in southern Europe. A long-legged centipede that moves at high speed, and can often be seen around houses.

HEXAPODA - insects

THYSANURA - bristletails

This small group is of interest since its members are well adapted to dry conditions and some manage to live on inhospitable recent lava.

Ctenolepisma longicaudata (Family Lepismatidae)

Not endemic. This species, or one of its close relatives, can often be seen momentarily as it scuttles out of sight, if one turns over rocks in barren places, including the upper parts of Mt.Teide.

ODONATA - dragonflies

There are 10 species of dragonflies on the island; these are discussed separately later, and an identification key is provided.

ORTHOPTERA - grasshoppers and crickets

Nearly 80 species of orthopterans are found in the Canaries, of which about one third are endemic: Tenerife has a good share of these.

Gryllus bimaculatus (Family Gryllidae - true crickets).

Not endemic. A robust glossy black cricket with two yellowish patches at the base of the wings. It is found mainly at fairly low levels in the north of the island, including agricultural areas.

Gryllomorpha canariensis (Family Gryllidae).

Canary endemic. This small, brown-grey, wingless cricket is widespread in the interior of the island, especially in the forests; however, we have also found it in Las Cañadas, both in a well at El Portillo and on a barren lava field.

Gryllotalpa africana (Family Gryllotalpidae - mole crickets).

Not endemic. This stout-bodied insect with enormously developed front legs spends most of its time tunnelling underground. It can be found - with luck - in a few damp places near the coast, including El Médano, San Andrés and Igueste.

Calliphona koenigi (Family Tettigoniidae - bush crickets or long-horned grasshoppers).

Tenerife endemic; occurs in north Tenerife. This is a powerful predatory insect with reduced wings; it has an adult body length of over 2.5cm (1in), but its green colour often makes it hard to see. The immatures can be found in flowers such as buttercups and the adults in trees.

Canariola nubigena (Family Tettigoniidae).

This was thought to be a Canary endemic genus but another representative has now been found in southern Spain. The species can be found on leaves in the laurel forest. It is small and wingless, with very long antennae (around four times body length); the female is brown and the male has a marbled grey ground colour with different colours superimposed.

Calliptamus plebeius (Family Acrididae - short-horned grasshoppers).

Canary endemic; common and widely distributed from near sea level up to Las Cañadas. Its body length is 2.5cm (1in) and it is very variable in colour.

Locusta migratoria, **Migratory Locust** (Family Acrididae).

Not endemic. This famous African species is a widespread resident in the lower parts of the island; it is often found in agricultural areas, but does not cause serious damage.

Oedipoda canariensis (Family Acrididae).

Canary endemic. This species is characterized by pale bluish hind wings. It is typical of open areas in the south and west of the island.

Schistocerca gregaria, **Desert Locust** (Family Acrididae).

Not endemic. This African locust reaches the island only sporadically, mainly in autumn, with warm winds from the Sahara. At long intervals there are major invasions, with serious damage to crops.

Sphingonotus willemsei (Family Acrididae).

Tenerife endemic. This is a ground-living species typical of the volcanic deserts of Las Cañadas, where it can become very abundant.

DERMAPTERA - earwigs

The earwigs of the Canaries are fascinating to animal geographers. The genus *Guanchia*, all of whose members are wingless, has four isolated species in northern Africa and southern Asia, but has undergone a spectacular adaptive radiation on the archipelago, giving rise to 10 endemic species, six of which occur on Tenerife. The genus *Anataelia* is also of great interest, with its closest relatives in China and Korea.

Anataelia canariensis (Family Pygidicraniidae).

Canary endemic. This wingless earwig is typical of the coastal spray zone along the north of the island, but some specimens - apparently of the same species - have recently been found in deep crevices in lava flows in Las Cañadas.

Guanchia cabrerae (Family Forficulidae).

Canary endemic. This vestigial-winged species, like the other members of the genus, is found only in woods in the north of the island.

DICTYOPTERA - cockroaches (Blattodea) and mantids (Mantodea)

The cockroaches of the Canaries are a mixture, with on the one hand species that have achieved worldwide (or at least very extensive) distribution with human aid (e.g. *Periplaneta americana*) and on the other hand species that are endemic to the Canaries or anyhow Macaronesia. The mantids are represented by seven species on Tenerife, and four of these are endemic.

Loboptera subterranea (Family Blattellidae).

Canary endemic. This blind cockroach, rich orange in colour, occurs in volcanic caves, including the Cueva de San Marcos (see EXCURSION No.21).

Phyllodromica (or *Arbiblatta*) *bivittata* (Family Blattellidae).

Canary endemic. This is a delicate cockroach, not associated with humans: it is especially characteristic of the pine and laurel forests, but also occurs under stones at up to 2500m (8200ft). Wings are present only in the male.

Mantis religiosa, **Praying Mantis** (Family Mantidae).

Not endemic. This species is widespread on the island, except at high levels and in the west; it is especially common on low vegetation near Los Rodeos airport. It is large - around 6cm (2.4in) - and is uniformly pale coloured, with fully developed wings.

Pseudoyersinia teydeana (Family Mantidae).

Tenerife endemic. This vestigial-winged species is found only in Las Cañadas, where it is always scarce. It lives on low vegetation but deposits its egg masses under stones.

Blepharopsis mendica (Family Mantidae).

Not endemic; also occurs in Africa. It is a dramatic spotted species up to 6cm (2.4in) long, which occurs mainly in dry areas near the coast.

ISOPTERA - termites

Only four species of termites have been found on the Canaries, and one of these is a human introduction.

Bifiditermes rogeriae (Family Kalotermitidae).

Canary endemic. This recently discovered species is an isolated outpost of a group that is otherwise confined to tropical Africa, southern Asia and Australia. It occurs in the rotting trunks of *Euphorbia* species.

THYSANOPTERA - thrips

These minute insects, many of them living in flowers, form a major part of the 'fallout' of aerially dispersing insects on the snow of Mt. Teide. The Canary forms have been well studied, and over 50 species are known from Tenerife, of which more than a quarter are endemic.

HEMIPTERA - homopteran and heteropteran bugs

Twenty families of homopteran bugs are found on the Canaries, though many of them are not particularly conspicuous. The leaf-hoppers are of interest because of the high proportion of endemic species and several endemic genera. The smaller forms - psyllids, whiteflies, aphids and coccids - include a number of important pests on cultivated plants on the island; a large proportion of the species seem to have been introduced by humans along with plants.

The 24 families of heteropteran bugs found on the Canaries are very diverse in their body form and habits, and include many colourful and conspicuous insects; in some families a high proportion of the species are endemic, and among the Miridae there are three endemic genera which have undergone adaptive radiation on the archipelago, with distinct species on different islands.

Dactylopius coccus (Family Dactylopiidae), **Cochineal Bug.**
  Not endemic. This is a small white woolly-looking bug whose body fluids are bright red. It was introduced from Mexico, along with its plant food **Prickly Pear**, for the production of cochineal, a deep red natural dye, which is still exported in a small way from Lanzarote.

Reduvius personatus (Family Reduviidae - assassin bugs).
  Not endemic: a Mediterranean species. A black ground-living predatory bug (which bites); 2cm (0.8in) long.

Velia lindbergi (Family Veliidae - water crickets).
  Endemic. This local species is very similar in appearance to its European relatives.

Notonecta canariensis (Family Notonectidae - backswimmers).
  Canary endemic. This is the only endemic among three species of backswimmers found on the island; the other two also occur in the Mediterranean region.

Euryderma ornatum (Family Pentatomidae).
  Not endemic. This conspicuous black and white bug can often be seen in large numbers on the crucifer *Descurainia bourgeauana* in Las Cañadas.

Aradus canariensis (Family Aradidae).

Canary endemic. A large blackish-brown bug that is normally restricted to the pine forests, although we have found a dispersing individual on snow on Mt.Teide.

NEUROPTERA - lacewings etc.

There are 26 species of neuropterans on Tenerife, and almost half of these are endemic to the archipelago. The ant-lions are likely to be of special interest to visitors from northern Europe; six species have been found on the island; the pits in which the larvae live can be found at any time of year, and in summer there is a chance of seeing the adults.

Myrmeleon alternans, **an ant-lion** (Family Myrmeleontidae).

Not endemic. This species is rather like a small dragonfly with lazy flight; its body has distinctive light and dark bands. It occurs both in sandy places in the lower zone and in Las Cañadas.

COLEOPTERA - beetles

This is the most diverse group of insects in the Canaries, with nearly 70 families represented; their evolution on the islands has fascinated entomologists for over a century. The ground-beetles (Carabidae) have been the focus of most attention; they include a large number of endemic species and genera, and are particularly associated with the laurel forest. However, several other groups have undergone notable radiations: for instance, the weevil genus *Laparocerus*, which is endemic to Macaronesia, has 60 species on the Canaries.

Calathus spp. (Family Carabidae).

A group of beetles that show very high diversity in the laurel forest, with no fewer than eleven species endemic to Tenerife. Some species live under the flaky bark of **Tejo.**

Carabus faustus (Family Carabidae).

Tenerife endemic. A ground-living beetle from the laurel forest; black but with brilliant green highlights.

Carabus interruptus (Family Carabidae).

Tenerife endemic. Smaller than preceding species, but similar in coloration. Typical of pine forest.

Eutrichopus martini (Family Carabidae).

Endemic blind beetle. Lives in caves.

Scarites buparius (Family Carabidae).

Not endemic. A large (3+cm) glossy black ground beetle. Found buried in sand dunes among roots of plants.

Cybister tripunctatus (Family Dytiscidae).

Not endemic. This is the largest (3+cm) species of freshwater diving beetle found on the island and is greenish-brown with a yellow band down the sides of the body.

Buprestis bertheloti (Family Buprestidae).

Endemic. The larvae of this species live in dead pine trunks, but adults can be seen flying at the tops of trees. It is very abundant in some years. Large, black, oval, with yellow pattern on elytra.

Temnochila pini (Family Ostomidae).

Canary endemic. Lives in large dead pine trunks: adults can be found under the bark. Large (2.5cm), brilliant black-blue.

Hegeter lateralis (Family Tenebrionidae).

Tenerife endemic. This is one of nine species of *Hegeter* found on Tenerife, seven of which occur only on this island; it is black with a pronounced bluish-grey "bloom" and can be found under rocks in the high mountain zone.

Nesotes conformis (Family Tenebrionidae).

Canary endemic. A heavily built black beetle that typically lives under the bark of the tree heath **Tejo**.

Pimelia radula (Family Tenebrionidae).

Tenerife endemic. A very large, black, broad-bodied beetle usually seen walking on the ground or trapped in old bottles; it is common in Las Cañadas.

Pimelia radula

Hesperophanes roridus (Family Cerambycidae - longhorn beetles).

> Canary endemic. A large beetle (more than 3cm long), mottled grey with a square head and strongly ridged elytra. Found in the high mountain zone.

Lepromoris gibba (Family Cerambycidae).

> Canary endemic. Large, dark grey. The larvae are found only in dead branches of **Cardón.**

Dicladispa occator (Hispa occator) (Family Chrysomelidae - leaf beetles).

> Canary endemic. A brown beetle with many spines, found in the pine forest where it feeds on species of *Cistus.*

Cionus variegatus (Family Curculionidae - weevils).

> Canary endemic. This is a variable species but it is basically creamy white and black. It feeds only on *Scrophularia* spp., and is common in Las Cañadas.

Cyphocleonus armitagei (Family Curculionidae).

> Tenerife endemic. Found only on *Argyranthemum teneriffae*. 2cm long, with black and white marks.

Laparocerus undatus (Family Curculionidae).

> Tenerife endemic. A chunky black weevil that can be found living under the bark of the tree heath **Tejo**.

LEPIDOPTERA - butterflies and moths

The Lepidoptera are very well represented in the archipelago, with a total of more than 500 species in nearly 50 different families. Over a quarter of the species are endemic. The butterflies and hawkmoths are discussed, together with identification keys, on later pages.

Dipsosphecia vulcania (Family Aegeriidae).

> Canary endemic. A day-flying, wasp-mimicking moth with transparent wings and dark body with pale bands.

Macaronesia (or *Dasichyra*) *fortunata* (Family Lymantriidae).

> A Canary endemic species, sometimes considered as the sole member of an endemic genus, *Macaronesia*, but sometimes placed in the widespread genus *Dasichyra*. A medium-sized hairy moth with mottled brown forewings and pale hindwings. The striking caterpillars, with tufts of coloured hairs, are a major pest on pine trees but can also be found on **Retama del Teide**.

Yponomeuta gigas (Family Yponomeutidae).

> Endemic. Larvae make communal webs over the endemic Canary Willow trees, and also on introduced poplars.

DIPTERA - flies

This order is represented by about 70 different families in the Canaries, with extremely diverse habits. In general few of the Diptera (or other insects) are troublesome to humans on Tenerife, although mosquitoes can be annoying at night in places near standing water.

Promachus vexator (Family Asilidae).

   Tenerife endemic. This very large predatory fly is most often seen along tracks in the forest zones.

Scaeva pyrastri (Family Syrphidae - hoverflies).

   Not endemic; widespread in the northern hemisphere. A conspicuous black and yellow fly often to be seen near the peak of Mt.Teide.

Tipula macquarti (Family Tipulidae - craneflies).

   Canary endemic. A large cranefly with vestigial wings, occurring in the laurel forest.

HYMENOPTERA - wasps, bees and ants

The more conspicuous groups of Hymenoptera (such as the social and solitary wasps and the bees) are fairly diverse on Tenerife and include a number of endemic species as well as some typical of northern Africa. There is also a wide variety of ants on the island, and again they include a number of endemic species. The parasitic Hymenoptera, many of which are tiny but brilliantly coloured, have been studied in much less detail: we have found large numbers stranded on snowfields on Mt. Teide, which included both well known widespread species and some apparently new endemic ones.

Bembix flavescens (Family Sphecidae).

   Not endemic. A solitary wasp typical of sandy areas behind beaches. Pale yellow and black.

Apis mellifera, **Honey Bee** (Family Apidae - bees).

   Introduced. There is a long tradition of bee-keeping on the island and some apiaries have over a hundred hives. A few bee keepers move their hives up to Las Cañadas in the early summer, and camp up there throughout the flowering season.

Bombus canariensis (B.terrestris canariensis) (Family Apidae).

   Endemic. This is the only bumblebee found in the Canaries, and is black with a white tail.

DRAGONFLIES

Only 10 species of dragonflies occur on Tenerife. Of these, one (the first in the list) is a damselfly while the other nine are typical dragonflies. These two groups are fairly easy to distinguish in the field: the damselflies are noticeably delicate, with very slender bodies, and at rest they hold their wings either vertically over the body or partly spread; the typical dragonflies are robust, strong-flying insects and always rest with wings outspread. All dragonflies have aquatic larvae and must have access to freshwater for breeding, but some of the larger species can be found several miles from their breeding places. They can be seen flying between April and October.

Anax imperator

The list below includes all the dragonflies known to occur on the island, and we also include a simple key to help with identification. For the purposes of the key and list "small" means with a body length of 30mm or less, "medium" 35mm-60mm and "large" 70mm-80mm. There is an excellent illustrated identification guide (in Spanish) to the dragonflies of the Canary Islands by Marcos Báez; this is available in good bookshops on the island (see APPENDIX, Books).

IDENTIFICATION KEY TO DRAGONFLIES

This key is hierarchical: at each step it is necessary to chose between the two entries that have the same number.

1. Small (body length 25-30mm), with wings much
    shorter than body, abdomen conspicuously
    slender, body green and black   *Ischnura saharensis*
1. Medium or large, abdomen robust
    2. Large (body length 70mm or more)
        3. Blue or green and black
            (see species accounts)   *Anax imperator* or *A. parthenope*
        3. Yellowish-brown, abdomen with dark
            dorsal stripe, and in male a blue mark near
            base   *Hemianax ephippiger*
2. Medium (body length 35-60mm)
    4. Abdomen conspicuously broad, red
        (male) or orange (female, young male)   *Crocothemis erythraea*
    4. Abdomen not conspicuously broad
        5. Red or pinkish
            6. Bright red, with black marks on sides
                of abdomen   male *Trithemis arteriosa*
            6. Pinkish or reddish, without distinct black
                marks on sides of abdomen (see species accounts)
                        male *Sympetrum fonscolombei* or *S. nigrifemu*r
        5. Blue, green, yellow or brown.
            7. Oblique white stripe on side of thorax,
                blue (male) or yellowish (female,
                young male)   *Orthetrum chrysostigma*
            7. Without oblique white stripe on side of
                thorax
                8. Black and yellow, strongly patterned
                    9. With metallic blue-green highlights
                                                Zygonyx torrida
                    9. Without metallic blue-green
                        highlights   female *Trithemis arteriosa*
                8. Yellow or yellowish-brown, not strongly
                    patterned (see species accounts)
                        female *Sympetrum fonscolombei* or *S. nigrifemur*

Family **COENAGRIIDAE**

Ischnura saharensis (I.elegans saharensis).

Not endemic; occurs in northern Africa, but not in Europe. (Has sometimes been considered a subspecies of *Ischnura elegans*.) Larvae probably develop in stagnant water. Small (body length 25-30mm); head and thorax green and black; abdomen strikingly slender and much longer than wings, dark above and greenish-yellow below, with a bright blue patch near the tip in the male.

Family **AESHNIDAE**

Anax imperator, **Emperor Dragonfly.**

Not endemic; occurs in south and central Europe, the Mediterranean region, Africa and parts of Asia. Larvae develop in stagnant water; adults are powerful fliers and are often seen far from water. One of the commonest of the dragonflies on the island. Large (body length 75-80mm); mainly blue or green; the male is distinguished by the colour of the abdomen, which is bright blue throughout its length, with distinct black dorsal marks; the female is similar but only the base of the abdomen is bright blue, the rest being green or blue-green: she could be confused with *A.parthenope*.

Anax parthenope, **Lesser Emperor Dragonfly.**

Not endemic; occurs in the Mediterranean region and Africa. Larvae develop in stagnant water; adults usually stay near water. Large (body length 75-79mm); similar to *Anax imperator*, but the abdomen is slightly shorter and darker in colour; the male has a small bright blue patch at the base of the abdomen, the rest being green or blackish-green; the female is generally darker.

Hemianax ephippiger, **Vagrant Emperor Dragonfly.**

Not endemic; occurs in the Iberian Peninsula, central Europe, and in desert areas of Africa and western Asia. Larvae sometimes develop in temporary waters; adults often fly far from water. Very rare in Tenerife. Large (body length 70mm), but distinctly smaller than the two *Anax* species; generally yellowish-brown, the abdomen with a dark dorsal line and in the male with a striking blue mark near the base.

Family **LIBELLULIDAE**

Crocothemis erythraea, **Scarlet Darter.**

Not endemic; occurs in central Europe, the Mediterranean region, Africa and into Asia. Larvae develop in either stagnant or running water; adults are found both close to water and far away. Medium (body length 40-50mm); in this species the whole body is conspicuously broad and the abdomen somewhat flattened above; the adult male is bright red all over, the young male and female golden yellowish; the abdomen has a dark dorsal line.

Orthetrum chrysostigma.

Not endemic; occurs in the south of the Iberian Peninsula and the whole of Africa. A dragonfly typical of dry areas, and adults are often found far from water. Medium (body length 50mm); the best character of this species is a conspicuous white stripe running from the base of the front wings downwards and forwards to the middle legs; the body colour is variable: adult males are mainly powder-blue, young males and females yellowish-brown; there are thin dark lines along the abdomen.

Sympetrum fonscolombei, **Red-veined Darter.**

Not endemic; occurs in west Europe, the Mediterranean region, Africa and into Asia. Larvae develop in stagnant, fresh or brackish water; adults stay in the vicinity. Medium (body length 35-41 mm); the male is generally pinkish, with the abdomen yellowish below, while the female is yellowish all over; in both sexes the spots at the front of the wingtips are yellow bordered with black, and tip of the abdomen has black dorsal marks (cf. *S.nigrifemur*, which is very similar).

Sympetrum nigrifemur.

Macaronesian endemic; also occurs on Gran Canaria and Madeira. Larvae generally develop in stagnant water; adults do not move far away. Very rare in Tenerife. Medium (body length 43-48mm); slightly larger than the very similar *S.fonscolombei*, and both sexes are darker in colour; the spots at the front of the wingtips are red or reddish-yellow and the abdomen lacks black dorsal marks at the tip.

Trithemis arteriosa.

Not endemic; occurs in Africa, especially in the tropics, but not in Europe. Larvae develop in stagnant or running water; adults often seen on stones nearby. Medium (body length 35-40mm); abdomen slender; entire body of the male is red, the abdomen brightest and with distinct black patches on the sides; in the female the entire body is strongly patterned black and yellow, and the yellow on the abdomen is mainly on top.

Zygonyx torrida.

Not endemic; occurs in the south of the Iberian Peninsula, Africa and into Asia. Larvae develop in swiftly running water; adults usually stay near water. Medium (body length 55-60mm), the largest of the libellulids on the island; abdomen patterned black and yellow with metallic blue-green highlights on the head and thorax.

217

BUTTERFLIES

This list includes all the 25 butterflies that are known to occur on Tenerife. Several species are endemic to the Canary Islands but entomologists disagree on the actual number: it is somewhere between three and seven. There is similar variation in opinion about the number that are endemic at the subspecific level. Over half of the species are southern ones that may be unfamiliar to many visitors; these are marked * in the systematic list. The English names, and the Latin ones which follow - with a few exceptions - are the ones used in A FIELD GUIDE TO THE BUTTERFLIES OF BRITAIN AND EUROPE (1983) by L.G.Higgins and N.D.Riley; alternative Latin names are given in brackets.

We have prepared a simple key to help with identification. In this key and in the individual descriptions we use "very small" to mean a wingspan of 20-34mm, "small" 32-44mm, "medium" 42-56mm, "large" 54-74mm and "very large" 72-120mm.

Monarch

IDENTIFICATION KEY TO BUTTERFLIES

This key is hierarchical: at each step it is necessary to choose between the two entries that have the same number.

1. Mainly white, yellow or pale orange
   2. Forewing with hardly any dark marks
      3. Strikingly yellow or pale orange, with one
         small spot on each wing   **Cleopatra**
      3. Greenish-white or pale yellow, with no
         spot on hind wing   **African Migrant**

2. Forewing with dark marks
    4. Bold dark margin to hindwing (as well as forewing), yellow or whitish   **Clouded Yellow**
    4. No dark margin to hindwing
        5. Mainly white and grey above, hindwing underside with distinct pattern of white on green
            6. White marks on hindwing underside form bars crossing the veins   **Green-striped White**
            6. White marks on hindwing underside form blotches between the veins   **Bath White**
        5. Mainly yellow or whitish, hindwing underside yellowish with grey dusting
            7. Medium or large, white or yellowish
                8. Medium, with small dark marks on forewing   **Small White**
                8. Large, with extensive dark tip to forewing   **Large White**
            7. Very small, yellow (rare, perhaps absent)   **Greenish Black-tip**

1. Brown, blue, coppery or brightly patterned
    9. Very small
        10. Yellow-brown, no strong pattern, tiny   **Lulworth Skipper**
        10. Not yellow-brown
            11. Strongly patterned, coppery or brilliant orange with dark brown
                12. Broad orange band near edges of both wings, above and below   **Brown Argus**
                12. Coppery triangle with black spots fills inner part of forewing, hindwing with small tail and with bright coppery band near edge   **Small Copper**
            11. Blue, blue-violet, dull coppery or brown above, without contrasting pattern
                13. With tail on hindwing, with a pair of dark spots at its base   **Long-tailed Blue**
                13. Without tail on hindwing
                    14. Underside of forewing orange-brown   **Canary Blue**
                    14. Underside of forewing grey with dark spots   **African Grass Blue**

9. Small to very large
    15. Strongly patterned with red or reddish-orange and black and white
        16. Very large or large, wings mainly orange with white-on-black border
            17. Veins on upperside very dark, hindwing with no black spots, or with a single tiny one; very large   **Monarch**
            17. Veins on upperside not noticeably dark, hind wing with several small black spots; large   **Plain Tiger**
        16. Small to large, wings with patches of red or orange, white, and black or brown
            18. Mainly black with vivid red and white marks
                19. Red band across forewing fairly straight and narrow   **Red Admiral**
                19. Red band across forewing broad with irregular invasions of black from behind
                    Indian Red Admiral
            18. Mainly orange, with dark brown and white marks
                20. Medium, underside of hindwing with five small eyespots   **Painted Lady**
                20. Small, underside of hindwing with two large eyespots (rare) **American Painted Lady**
    15. Relatively uniform brown, orange or both, sometimes with many black marks
        21. Brown or brown and orange with a few eyespots
            22. Forewings each with two dark eyespots, showing most clearly on underside; blackish-brown   **Canary Grayling**
            22. Forewings each with a single dark eyespot
                23. Forewing with chequered pattern, upperside of hindwing with several eyespots on orange patches   **Canary Speckled Wood**
                23. Forewing with poorly defined yellowish or orange patches, upperside of hindwing without eyespots   **Meadow Brown**
        21. Brownish or orange-brown with many black marks
            24. Large, pale greenish-brown with bold black marks
                    Cardinal
            24. Small, rich orange-brown with many black spots (rare)
                    Queen of Spain Fritillary

Family **PIERIDAE**. All have white or yellowish-white wings, generally with some dark markings.

Large White, *Pieris brassicae cheiranthi (P.c.cheiranthi)*, **Blanca de la Col.**

Canary endemic subspecies of a species which occurs in Europe and northern Africa; subspecies also occurs on La Palma and Gomera. (It is sometimes considered to be a separate species.) Patchily distributed; found in the lower and forest zones from sea-level to about 1200m (3950ft), especially in the north. Large (wingspan 58-64mm); mainly white and pale yellow, the forewing with broad dark tip above, and with two conspicuous merging dark marks (on both surfaces in female and on lower surface only in male). It is larger than the form found in northwest Europe and the dark spots are larger and unite. Larval foodplants Cruciferae.

Small White, *Artogeia rapae (Pieris rapae)*, **Blanquita de la Col.**

Not endemic; occurs in Europe and northern Africa and on all the Canary Islands. Common; found throughout the island, especially in cultivated areas. Medium (wingspan 42-50mm); predominantly white on upperside; forewing with small black tip and two small, well-separated black spots: underside of hindwing yellowish, dusted with grey. Larval foodplants Cruciferae and Resedaceae.

*Bath White, *Pontia daplidice (Pieris daplidice, Leucochloe daplidice)*, **Blanquiverdosa.**

Not endemic; occurs in central and southern Europe and northern Africa and on all the Canary Islands. Found throughout the island. Small (wingspan 38-44mm); predominantly white and pale grey above with strong dark markings mainly towards the tip of the forewing; underside of hindwing dark green with whitish blotches mainly confined between the veins (cf. **Green-striped White**). Larval foodplants Cruciferae and Resedaceae.

*Green-striped White, *Euchloe belemia eversi (E.b.hesperidum)*, **Blanca Verdirayada.**

Tenerife endemic subspecies of a species which occurs in southern Europe and northern Africa; another subspecies occurs on Gran Canaria and Fuerteventura. (Some authors do not separate the Tenerife form.) Found in open areas in the pine forest and in the high mountain zone up to at least 2500m (9000ft). Small (wingspan 32-36mm); similar to **Bath White** but smaller and with the white marks on the green underside of the hindwing in the form of bars crossing the veins. Larval food plant *Descurainia bourgeauana*.

*Greenish Black-tip, *Elphinstonia charlonia (Euchloe charlonia)*, **Puntaparda Verdosa.**

Not endemic; occurs in northern Africa and on Fuerteventura and Lanzarote. Has been recorded from Tenerife, but there are no recent records; most likely on dry hilltops in the south in spring. Very small (wingspan 28-32mm); pale yellow above, forewing with dark tip and single conspicuous dark mark near front edge; yellowish-green below.

*African Migrant, *Catopsilia florella*, **Migradora Africana.**

Not endemic; occurs in Africa south of the Sahara and now also on Gomera and Gran Canaria. Found in the lower zone. Large (wingspan 54-58mm); pale greenish-white, but some females are yellow-buff; forewing with a small dark dot and in female the front edge is dark; distinguishable from the other "whites" at a distance by its much faster and more erratic flight. First recorded in Tenerife in 1966, after introduction of its foodplant *Cassia didymobotrya.*

Clouded Yellow, *Colias crocea*, **Colias Común.**

Not endemic; occurs in central and southern Europe and northern Africa and on all the Canary Islands. Widespread and abundant in the lower and forest zones. Medium (wingspan 42-50mm); orange-yellow (yellowish-white in some females) with bold dark margin to both forewing and hindwing (with pale spots in it in the female) and a conspicuous dark spot on forewing. Larval foodplants Leguminosae.

*Cleopatra, *Gonepteryx cleopatra cleobule (G.cleobule)*, **Cleopatra.**

Canary endemic subspecies of a species which occurs in southern Europe and northern Africa; this subspecies also occurs on Gomera (it is sometimes considered to be a separate species) and another subspecies occurs on La Palma. Found in rough ground from about 600-1800m (2000-6000ft), especially in the laurel forest. Large (wingspan 56-60mm); bright greenish-yellow, with upperside of forewing orange; a small orange spot near the centre of each wing. Larval foodplants on the island not known.

Family **LYCAENIDAE**. All very small and with blue, coppery or brown wings.

Small Copper, *Lycaena phlaeas*, **Manto Bicolor.**

Not endemic; occurs in Europe and northern Africa and on all the Canary Islands. Common throughout the island, though rare in high mountain zone. Very small (wingspan 24-32mm); coppery coloured forewing with dark brown edge and black spots; small tail and bright coppery band on hindwing; forewing underside orange with black spots. Larval foodplants docks.

*Long-tailed Blue, *Lampides boeticus*, **Canela Estriado.**

Not endemic; occurs in central and southern Europe and northern Africa and on all the Canary Islands. It is found locally in the lower zone. Very small (wingspan 29-34mm); distinctly hairy appearance; male is violet-blue above; the hindwing has a small tail with a pair of prominent dark spots at its base, showing above and below; the underside in general is pale grey-brown, mottled and barred with white; female is mainly brown above but with some violet-blue near bases of wings. Larvae live inside the seed pods of legumes.

*Canary Blue, *Cyclyrius webbianus*, **Manto de las Canarias.**

Canary endemic species; also occurs on the three westernmost Canary Islands and Gran Canaria. Widespread throughout the island, and found on open ground from sea level to 3000m (10000ft) on Teide. Very small (wingspan 26-30mm); both fore and hindwing are violet-blue above in the male and golden brown in the female, with the border somewhat darker and the extreme edge pale; the underside of the forewing in both is orange-brown and the hindwing is mottled brown with a stepped white bar. Larval foodplants Leguminosae, including *Teline canariensis, Spartocytisus supranubius, Adenocarpus* spp. and *Lotus* spp.

*African Grass Blue, *Zizeeria knysna*, **Niña Esmaltada Menor.**

Not endemic; occurs in the extreme south of Europe, Africa and on all the Canary Islands. Common at low altitudes in moist places. Very small (wingspan 22-25mm); male is dark violet-blue above and female is brown above with some blue at base of wings; both are silvery-grey with dark spots below. Larval foodplants include Leguminosae.

Brown Argus, *Aricia agestis cramera (A.cramera)*, **Morena Serrana.**

Not endemic; the subspecies occurs in the Iberian peninsula, Balearic Islands, northwest Africa and on Gomera, La Palma and Gran Canaria; the species also occurs in Europe and into Asia. (The subspecies is sometimes considered to be a separate species.) Found in some parts of the laurel and pine forest regions up to about 2000m (6600ft). Very small (wingspan 22-26mm); dark brown above and pale and spotted below, both wings having a row of brilliant orange markings near the edge on both surfaces. This subspecies is brighter coloured than the widespread form. Larval food-plants Leguminosae.

Family **NYMPHALIDAE.** Small to large butterflies, usually with bright colours and strong patterns.

Red Admiral, *Vanessa atalanta*, **Vanesa Atlanta.**

Not endemic; occurs in Europe and northern Africa and on all the Canary Islands except Lanzarote. Fairly widely distributed in the lower and forest zones but not in the extreme south; also found in gardens. Medium (wingspan 48-54mm); predominantly velvety black with white spots near tip of forewing and a brilliant red band across forewing, showing on both surfaces - this band is fairly straight and narrow (cf. **Indian Red Admiral**); red border to hindwing on upper surface. Larval foodplants nettles (*Urtica* sp.) or thistles.

*Indian Red Admiral, *Vanessa indica callirhoe (V.i.vulcania, V.vulcania)*, **Vanesa India.**

Macaronesia endemic subspecies of a species which also occurs in India and eastern Asia; the subspecies occurs on all the Canary Islands except Lanzarote, and on Madeira. Found in the laurel forest, a few parts of the lower zone and pine forest, and also in gardens. Medium or large (wingspan 48-60mm); like the **Red Admiral** but the red band on the forewing is broader and with irregular invasions of black from behind. Larval foodplants not known.

Painted Lady, *Cynthia cardui*, **Cardera, Vanesa de los Cardos.**

Not endemic; occurs in Africa and on all the Canary Islands and in Europe in summer. Widespread throughout most of the island; sometimes arives with a southeast wind from Africa. Medium (wingspan 46-52mm); pale orange-brown, with complex pattern of black and white markings; especially dark near tip of forewing. Underside of hindwing has five small eyespots (cf. **American Painted Lady**). Larval food plants normally thistles and nettles.

*American Painted Lady, *Cynthia virginiensis*, **Vanesa Americana.**

Not endemic; occurs in North America, Madeira and also on the westernmost Canary Islands and Gran Canaria. Rare; found in the laurel forest and in gardens; occasionally in high mountain zone. Small (wingspan 36-46mm); similar to **Painted Lady**, but smaller, with more rounded wings; underside of hindwing has two very large eye-spots. Larval foodplants various Compositae.

*Cardinal, *Pandoriana pandora (P.pandora seitzi)*, **Pandora.**

Endemic status disputed; the species occurs in southern Europe and northern Africa and on La Palma and Gomera (the Canary Island form is sometimes considered to belong to a distinct subspecies). Not common, but sometimes found in pine forest and mixed forest. Large (wingspan 56-70mm); pale greenish-brown above, with many bold black marks; lower surface of forewing is rose-pink at base, while the hindwing is pale green with silvery bars. It has a gliding flight. Larval foodplants violets (*Viola* sp.).

*Queen of Spain Fritillary, *Issoria lathonia*, *Sofía.*

Not endemic; occurs in Europe and northern Africa and on Gomera and La Palma; a species of continental Europe which only rarely gets into south Britain. Rare, but occurs in the laurel forest of Monte del Agua. Small (wingspan 36-42mm); orange-brown upper surface with many dark spots, lower surface of hindwing with large silvery markings. Larval foodplants violets (*Viola* sp.).

Family **SATYRIDAE.**

Medium to large butterflies, typical of grassland or open woodland; usually brown.

*Canary Grayling, *Hipparchia wyssii wyssii (Pseudotergumia wyssii wyssii)*, *Sátiro Moreno de las Canarias.*

Canary endemic subspecies of a Canary endemic species; the subspecies also occurs on Gran Canaria (other subspecies occur on Hierro, Gomera and La Palma). Widespread in the pine forest, and especially common between about 1400 and 1600m (4600 and 5250ft) in the south; also occurs in the high mountain zone. Medium-large (wingspan 50-60mm); upper surface of wings uniformly blackish-brown with very narrow white fringe and two dark eyespots (more conspicuous on lower surface); a few small white marks on forewing. Larval foodplants not recorded, but probably grasses.

Meadow Brown, *Maniola jurtina hispulla (M.j.fortunata, M.hispulla fortunata)*, *Loba.*

Not endemic; occurs in Europe and northern Africa and also on all the three westernmost Canary Islands and Gran Canaria. (The form *hispulla*, in the western Mediterranean and Canaries, is often considered as a species separate from *jurtina*.) Found in open areas in the forest zones. Medium (wingspan 44-54mm); greyish-brown with areas of orange-brown (almost absent in males on upperside) and a single black eyespot on the forewing (obscure on the upperside in males); larger than the northern form and with larger eyespot and somewhat brighter colouring. Larval foodplants grasses.

*Canary Speckled Wood, *Pararge xiphioides (P.aegeria xiphioides)*, *Maculada de las Canarias.*

Canary endemic species; also occurs on Gomera, La Palma and Gran Canaria. (It is sometimes considered a subspecies of the ordinary Speckled Wood.) Common in the forest zones and parts of the lower zone, and also occurs in gardens. Small (wingspan 38-44mm); brown forewing with many orange and yellow patches and one small eyespot; hindwing with a row of small eyespots in orange patches. Larval food plants grasses.

Family **DANAIDAE.** Large and brightly coloured butterflies capable of long-distance flights.

Monarch (Milkweed)**, *Danaus plexippus*, **Monarca.

Not endemic; occurs throughout the Americas and parts of Asia and on all the Canary Islands except Fuerteventura and Lanzarote; it colonized the Canaries in 1880, presumably helped across the Atlantic by exceptional westerly winds. It is now found in the lower zone, but seems to be common only in gardens; it has been recorded from Las Cañadas. Very large (wingspan 80-120mm); wings orange with very dark veins; small white spots in dark band around edge of wings and also at tip of forewing. Both adult butterflies and the caterpillars (coloured yellow, black and white) are distasteful as the result of feeding on *Asclepias curassavica*, which contains poisonous substances; it also feeds on *Ceropegia* spp., which are also Asclepiadaceae. The bright patterns are considered to be a case of "warning coloration".

Plain Tiger**, *Danaus chrysippus*, **Mariposa Tigre.

Endemic status disputed; occurs in Africa south of the Atlas mountains and on all the Canary islands except Hierro and Lanzarote (the Canary Island form is sometimes considered as an endemic subspecies). Found locally in the lower zone, mainly in gardens. Large (wingspan 62-74mm); smaller than **Monarch** but fairly similar; veins not so conspicuous but with several dark spots in the hindwing. The larvae feed on varous Asclepiadaceae, including *Asclepias curassavica, Gomphocarpus fruticosus* and *Ceropegia* spp.

Family **HESPERIIDAE.** A family showing a number of structural differences from other butterflies; often rest with the wings closed down over the body, like moths.

Lulworth Skipper**, *Thymelicus acteon christi (T.christi)*, **Dorada Oscura.

Canary endemic subspecies of a species that occurs in central and southern Europe and northern Africa; the subspecies also occurs on Gomera, La Palma and Gran Canaria (it is sometimes considered a separate species). Found locally in the lower and forest zones in open places and on banks, up to about 1800m (6000ft). Very small (wingspan 20-24mm) and rather moth-like, but with "clubbed" antennae; coppery orange-brown with no conspicuous dark markings. Larval foodplants grasses.

HAWKMOTHS

This list includes the six hawkmoths that are known to occur on Tenerife; all of them are also found in Europe. The names are those used in THE MOTHS AND BUTTERFLIES OF GREAT BRITAIN AND IRELAND, ed. John Heath (1979); alternative Latin names are given in brackets. In the simple identification key and in the individual descriptions we use "medium" to mean a wingspan of 42-56mm, "large 54-74mm and "very large" 72-120mm.

IDENTIFICATION KEY TO HAWKMOTHS

1. Forewing mottled grey-brown without longitudinal pattern; no pink on hindwing
    2. Medium, day-flying, hindwing yellow and brown
        Hummingbird Hawkmoth
    2. Very large
        3. Yellow bands on abdomen, hindwing yellow with dark bands **Death's Head Hawkmoth**
        3. Pink bands on abdomen, hindwing grey with faint dark bands **Convolvulus Hawkmoth**
1. Longitudinal pattern on forewing, pink on hindwing
    4. Abdomen not strongly banded, hindwing with pink patch close to body, forewing with narrow silvery stripe **Silver-striped Hawkmoth**
    4. Abdomen with light and dark bands, hindwing with broad pink band
        5. Forewing with distinct pale stripe along centre, crossed by white veins, top of thorax with two white stripes **Striped Hawkmoth**
        5. Forewing has broad pale area with dark blotches near front edge, no white veins, thorax with white only at sides **Spurge Hawkmoth**

Family **SPHINGIDAE**
Convolvulus Hawkmoth, *Agrius convolvuli (Herse convolvuli)*.
Not endemic; widespread in the Old World. Found mainly in the lower zone and laurel forest. Very large (wingspan 89-120mm); forewing,

hindwing, head and thorax grey mottled with whitish; abdomen with pink, white and black bands and broad dark dorsal stripe. Caterpillar apple-green or purplish, with oblique yellowish stripes on the sides; brownish tail horn. Larval foodplant typically *Convolvulus*, but can also be found on grape vines.

Death's Head Hawkmoth, *Acherontia atropos.*

Not endemic; occurs throughout Europe and the Middle East and Africa. Found mainly in the lower zone. Very large (102-135mm); forewing mottled brown with one tiny white spot; hindwing yellow with two dark bars; thorax brown with yellow "death's head" pattern; abdomen with yellow and dark brown bands, and broad dorsal grey stripe. Caterpillar up to 12.5cm (5in) long; greenish with 7 oblique purplish sidestripes edged with yellow; tail horn rough, yellowish. Larval foodplants mainly Solanaceae.

Hummingbird Hawkmoth, *Macroglossum stellatarum (Macroglossa stellatarum).*

Not endemic; occurs in Europe and north Africa. Found throughout the island. Medium (wingspan 50-58mm); forewing brownish with a few dark transverse lines; hindwing yellow, edged with reddish-brown; abdomen stumpy, brownish with pale patches on the sides. This hawkmoth is a day-flyer and darts from flower to flower hovering in front of them while it feeds. Caterpillar green or brownish with white dots and pale longitudinal lines; tail horn bluish. Larval foodplants Rubiaceae.

Spurge Hawkmoth, *Hyles euphorbiae tithymali (Celerio euphorbiae tithymali).*

Endemic subspecies of a species that is widespread in Europe and eastwards to India. Found throughout the island except the high mountain zone. Large (wingspan 64-77mm); forewing greyish-brown, slightly pinkish and with bold dark brown markings and a large yellowish area with irregular edges along centre; hindwing with whitish patch near body, then broad pink band bordered with black; thorax brown above, white at sides; tapering abdomen with blackish bands and brown dorsal stripe. Caterpillar blackish with crimson dorsal stripe, yellowish lateral stripes and a series of bold round spots; tail horn red. Larval foodplants *Euphorbia* species.

Striped Hawkmoth, *Hyles lineata (Celerio lineata, Celerio livonica).*

Not endemic; worldwide distribution. Found throughout most of the island except the high mountain zone. Very large (wingspan 78-90mm); forewing mainly dark brown with pale edge, broad pale stripe down centre and diagnostic white veins; hindwing whitish near body, pink further out, both bordered with black; thorax brown above with two white stripes; pointed abdomen with black bands broken with pale spots. Caterpillar green to black with yellow dorsal and lateral stripes and a series of bold round spots; tail horn red. Larval foodplants include Rubiaceae, docks and grapevine.

Silver-striped Hawkmoth, *Hippotion celerio.*

Not endemic; widespread in southern Europe and Asia and the rest of the Old World. Found mainly in the lower zone and laurel forest. Large (72-80mm); forewing pale brownish with narrow silvery-white longitudinal stripe; hindwing with pink patch close to body, pale cells surrounded by black further out; abdomen tapered rather than stumpy, brown with no bands but with two thin white dorsal stripes. Caterpillar brown or greenish, with narrow head end, two large eyespots a little way back and two smaller spots behind them; tail horn rough, brown. Larval foodplants include grapevine and Rubiaceae.

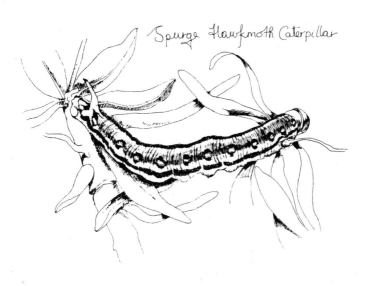

Spurge Hawkmoth Caterpillar

SELECTED MARINE INVERTEBRATES

This list includes only those marine invertebrates that are mentioned in the text.

PORIFERA - sponges

Clathrina coriacea, **sponge.**
A small bright red sponge that can be found growing on the lower surface of rocks in pools in the intertidal zone.

COELENTERATA - jellyfish, anemones, etc.

Anemonia sulcata, **anemone.**
A pale green anemone that is common in rock pools.
Balanophyllia regia, **coral.**
A tiny red or yellow coral that can be seen in the intertidal zone.
Velella spirans, **By-the-wind-sailor,** *Vela.*
This colonial animal is in the form of an oval blue disc with a diagonally set sail on the top and tentacles below. It is a warm water oceanic species that can sometimes be found washed up on beaches.
Physalia physalis, **Portuguese Man-'o-war,** *Fragata Portuguesa.*
An oval gas-filled bladder about 15cm (6in) long, forming a sail, from which hang a mass of tentacles equipped with powerful stinging cells. An oceanic species that can sometimes be seen in coastal waters or on beaches, especially in late winter.

ANNELIDA

POLYCHAETA - bristle worms

Eulalia viridis, **paddle worm.**
A slender green worm up to 10cm (4in) long, that can swim as well as walk; it is found among algae in the intertidal zone. This species also occurs in western Europe.

CRUSTACEA

CIRRIPEDIA - barnacles

Chthamalus stellatus, **acorn barnacle.**
This species also occurs on the coasts of Europe as far north as Britain. In many parts of the coast of Tenerife the rock surfaces high in the intertidal zone are covered by dense populations of these barnacles.

DECAPODA - crabs, shrimps, etc.

Clibanarius aequabilis, **hermit crab.**

Many of the mollusc shells in rock pools prove to contain these gaily coloured hermit crabs rather than the legitimate owners.

Pachygrapsus sp, **crab.**

A tiny crab that lives in holes in rock pools.

Palaemon elegans, **shrimp,** *Camarón, Quisquilla.*

There is a local fishery for shrimps, and they are often cooked and served in bars near ports.

Percnon planissimum, **crab.**

A flat crab that seems to slither along the rocks.

MOLLUSCA

GASTROPODA - marine snails, sea slugs, etc.

Aplysia dactylomela, **sea hare,** *Conejo de Mar.*

A slug-like mollusc up to 1ft (30cm) long; very variable in colour but generally pale. It is found from the Cape Verde Islands as far north as the Canaries.

Glaucus atlanticus, **nudibranch, sea slug.**

A purple, flattened mollusc with extraordinary branched projections from the sides of its body. It is oceanic, floating at the sea surface, where it feeds on coelenterates.

Haliotis coccinea canariensis, **abalone,** *Oreja de Mar.*

Macaronesia endemic. This species can be found under rocks in pools; it has a gently curved oval shell with a row of holes along one edge, and is up to 7cm (3in) long.

Littorina striata, **winkle.**

Macaronesia endemic: a warm-water relative of the winkles of western Europe. It occurs high up on the shore, and is dark-coloured, up to 1cm long, and with a pointed shell.

Osilinus atratus, **topshell.**

Macaronesia endemic. A fairly low-spired patterned snail, occurring high up on the shore.

[*Patella candei,* **limpet,** *Lapa.*

This large limpet may now be extinct on Tenerife, but was evidently eaten by the early human inhabitants.]

Patella piperata, **limpet,** *Lapa.*

Macaronesia endemic. A close relative of the species found on European coasts. These limpets occur high up on the shore, and are often abundant enough to be collected and served in local restaurants as lapas.

Thais haemastoma, **whelk, *Purpura*.**

A large (6cm - 2.4iñ) snail with a knobbly shell and orange aperture more than half the height of the shell; it is carnivorous, feeding on barnacles and molluscs. The species is widespread in the Mediterranean and Atlantic.

Vermetus gigas, **worm-shell.**

Member of an unusual group of sedentary gastropod molluscs which look like curled up calcareous worms.

CEPHALOPODA - octopuses, squids, etc.

Octopus vulgaris, **octopus, *Pulpo*.**

This species also occurs in Europe, although not in the north. It is abundant in the lower part of the intertidal zone, and is much hunted for food.

ECHINODERMATA - sea urchins etc.

Arbacia lixula, **sea urchin.**

This species occurs low down on the shore, in sheltered places; it is black and feeds on calcareous algae.

Paracentrotus lividus, **sea urchin.**

This urchin also occurs in western Europe; it varies in colour from purple to greenish black, and can sometimes be found wedged into crevices in rock pools.

Ophioderma longicauda, **brittle star.**

A black brittle star found under rocks in pools low down in the intertidal zone.

Octopus

APPENDIX

How to get to the islands

If you intend to go for a short stay in Tenerife, undoubtedly the most economic way is to go with one of the numerous charter companies. They fly to the international airport which is in the south of the island, and several of them offer flight-only holidays. Unless transport from the airport to your hotel or appartment is provided, be prepared for a fairly expensive taxi journey. There are, however, fairly frequent buses to Santa Cruz which should help if your destination is in the north. Car hire is sometimes cheaper in the towns than at the airport.

A more leisurely way to reach the islands is to go by car and boat. From Britain you can take the ferry to Santander, and drive south to Cádiz. From Cádiz there is a ferry to the islands; the trip would take five or six days from south Britain.

When to go

If plants are your main interest, then the best time to go is between April and July. During this period you will be certain to see many of the endemic Canary species of plants in flower. The weather can be very hot from July to September, but at least you will be unlikely to have any rain then; most of the rain falls between October and February. During the winter months it can be very cold indeed in the high parts of the island, especially at night; the coastal areas, however, are never very cold.

Where to stay

If you go to the islands primarily for the sun, it is worth remembering that the north of the island lies under cloud for much of the time from February to October; there is a good chance of clear skies from November to January. The south of the island is much sunnier all the year round.

If you book a holiday at a tourist hotel in the south you may well find that you are in an isolated tourist resort far from public transport, and you may not be able to get much information about it: bear this in mind if you intend to use the buses. From the transport point of view the best places to stay are Puerto de la Cruz or Santa Cruz. If you are not worried about booking accommodation in advance you can get a list of hotels and also '*hostales*' (which generally do not provide meals) from the Spanish Tourist office (57/58 St.James Street, London SW1 A1LD) and work out a holiday that will take you to more remote parts of the island. You should be prepared for fairly simple accommodation if you do this.

Camping

There are no official camp sites in Tenerife. Camping is not allowed in the Teide National Park or any of the forest areas. There are, however, several coastal areas where people do camp; no sanitary facilities are provided.

What to wear

Be prepared for extremes of temperature, and also for wind and rain. Never go on a trip without an extra layer of clothes, even if this seems ridiculous on a sunny morning. If you go up high you will find that the temperature drops very sharply at sunset; in winter you will need warm hill-walking gear even during the day. Several of the places described in this book require good walking shoes, or preferably hiking boots. Have a waterproof/windproof jacket with you. High altitude lip cream and suntan lotion are advisable as a protection against the wind and sun.

Maps

The best general map is put out by the Cabildo Insular de Tenerife, but it has recently been out of print. If you buy one, check that the scale is right: in one edition the scale is 1:100000, as stated, but the printed scale line at the bottom of the map is graduated from 0 to 20km, whereas it should be 0 to 10km, so all distances are half what the scale would indicate. Also, do not try to drive to Vilaflor along the white road from near the observatory at Izaña: most of it is passable, though rough, but there is no way across the Barranco del Rio! More detailed military maps are available from No 3. Avenida 25 de Julio, Santa Cruz (the office is open from 09.00-14.00 and you must take your passport with you when you go) and also by mail from Edward Stanford (12-14 Long Acre, London). Eight of these maps cover the island at a scale of 1:50000.

Road maps are available in shops, kiosks and petrol stations, but these are of little help for anyone interested in getting away from the main roads. We have come across some tourist maps published in Britain, but have yet to find one that does not perpetuate mistakes made in earlier maps. In contrast, the detailed walk maps in LANDSCAPES OF TENERIFE (see BOOKS) are exceedingly good.

Botanic gardens

Although the official title of the Jardín Botánico refers to Orotava this garden is on the outskirts of Puerto de la Cruz and not in the town of Orotava. It is not primarily concerned with the native Canary plants, but a visit there may well help you sort out the identification of many of the palms and other introduced ornamental plants that you will see growing in the urban areas.

Museums

The Natural History Museum of Tenerife in Santa Cruz (Museo Insular de Ciencias Naturales) has recently been expanded to include the archaeology and anthropology museums. These are now all housed in the old civil hospital, a fine building on the corner of Avenida Bravo Murillo and of Calle San Sebastian, not far from the Plaza de España; an extensive renovation programme has recently been in progress. Here you will find an aquarium, terrarium and departments for entomology, botany, minerals, fossils, natural history, archaeology, anthropology, geology and vulcanology. The emphasis is on the Canary Islands and Macaronesia.

Bookshops

Many bookshops, and also some more general shops that sell books, have a special section on the Canaries. Although there are bookshops in Los Cristianos, Playa de Las Americas and Puerto de la Cruz, we have found the best selection of books on the natural history of the Canaries in Librería Lemus (La Laguna), Librería Goya (Santa Cruz) and Teide National Park Visitor Centre in Las Cañadas. For anyone wanting to visit several bookshops, the university town of La Laguna would be the place to start.

BOOKS

The travel shelves of any large bookshop in Britain should produce visitor guides to Tenerife. However these have little useful information on the natural history. Apart from WILDFLOWERS OF THE CANARY IS-LANDS by David and Zoë Bramwell, most of the information in English concerning the natural history of the island is to be found in a number of old books and various specialized journals. However, in recent years, several excellent books in Spanish have appeared. We have divided the following list into two sections. The first part deals with books (mostly in Spanish) which are either entirely or partially concerned with various aspects of the natural history of the Canary Islands. The second section lists other more general books that we have referred to in the text.

Books on Canary Islands natural history

The following list is only a selection; a visit to the best bookshops on the island might well provide more titles. Several of the books, even of recent date, are quite hard to obtain.

ATLAS BASICO DE CANARIAS, 1980. Editorial Interinsular Canaria, Santa Cruz de Tenerife.

This useful atlas includes maps showing geology, climate, vegetation, soil types and agriculture.

Araña, V. and Carracedo, J.C. 1978. *CANARIAN VOLCANOES* 1. TENERIFE. Editorial Rueda, Madrid.

This book is in both Spanish and English. The translation is somewhat idiosyncratic but it is a useful introduction to the volcanic geology of the island.

Bacallado Aranega, J.J. (editor). 1984. *FAUNA MARINA Y TERRESTRE DEL ARCHIPIELAGO CANARIO*. Editorial Regional Canarias, Las Palmas de Gran Canaria.

A comprehensive account of the animals of the islands, by professional zoologists; packed with information and with a great many colour photographs, line drawings, tables and diagrams, and bibliographies for the various animal groups. In Spanish.

Báez, M. 1986. *LAS LIBELULAS DE LAS ISLAS CANARIAS*. Aula de Cultura, Cabildo Insular de Tenerife.

An account of the dragonflies of the Canary Islands, illustrated with colour photographs. In Spanish.

Bannerman, D.A. 1922. THE CANARY ISLANDS - THEIR HISTORY, NATURAL HISTORY AND SCENERY. Gurney and Jackson, London.

Now only available occasionally from second-hand dealers. A good historical account of the travels of a famous ornithologist.

Bannerman, D.A. 1963. BIRDS OF THE ATLANTIC ISLANDS. Vol.I: A HISTORY OF THE BIRDS OF THE CANARY ISLANDS AND THE SALVAGES. Oliver and Boyd, Edinburgh.

Now out of print but available in specialist libraries. A very complete account of the birds.

Bramwell, D. and Bramwell, Z.I. 1974. WILD FLOWERS OF THE CANARY ISLANDS. Stanley Thornes, Cheltenham.

This book, amply illustrated with colour photographs and line drawings, is a must for any flower lover. It is a fairly technical book that concentrates on the endemic plants of the Canaries; it omits most Mediterranean plants, of which there are many on the island.

Bramwell, D. and Bramwell, Z.I. 1987. *HISTORIA NATURAL DE LAS ISLAS CANARIAS*. Editorial Rueda, Madrid.
An illustrated guide to a selection of plants and animals. In Spanish.

Ceballos L. and Ortuño, F. 1976. *VEGETACION Y FLORA FORESTAL DE LAS CANARIAS OCCIDENTALES*. Cabildo Insular de Tenerife.
A comprehensive account of the vegetation of the western Canary Islands, with beautiful photographs. In Spanish.

Fernández, J.M. 1978. *LOS LEPIDOPTEROS DIURNOS DE LAS ISLAS CANARIAS*. Aula de Cultura, Cabildo Insular de Tenerife.
An account of the butterflies of the Canary Islands, illustrated with colour photographs. In Spanish.

Gobierno de Canarias. 1985. *GUIA DE PECES, CRUSTACEOS Y MOLUSCOS DE INTERES COMERCIAL DEL ARCHIPIELAGO CANARIO*.
Line drawings and notes on fish, molluscs and crustaceans of commercial importance. In Spanish.

González Henríquez, M.N. et al. (editors). 1986. *FLORA Y VEGETACION DEL ARCHIPIELAGO CANARIO*. Editorial Regional Canarias, Las Palmas de Gran Canaria.
A companion volume to *FAUNA MARINA Y TERRESTRE DEL ARCHIPIELAGO CANARIO*, described above.

Hansen, A. and Sunding, P. 1985. FLORA OF MACARONESIA: CHECKLIST OF VASCULAR PLANTS, 3rd revised edition. Sommerfeltia No.1. Botanical Garden and Museum, University of Oslo, Norway. Not available in bookshops. In English.

Hernández Hernández, P. (editor). 1978. *NATURA Y CULTURA DE LAS ISLAS CANARIAS*. Printed by Litografía A. Romero, Tenerife.
This is a concise encyclopaedic book with a little about everything on the islands. In Spanish, but English edition in preparation.

Manuel Moreno, J. 1988. *GUIA DE LAS AVES DE LAS ISLAS CANARIAS*. Editorial Interinsular Canaria, Tenerife.
An identification guide to both resident and non-resident birds.

Martín Hidalgo, A. 1987. *ATLAS DE LAS AVES NIDIFICANTES EN LA ISLA DE TENERIFE*. Instituto de Estudios Canarios, La Laguna.
A comprehensive and up-to-date book on the breeding biology and distribution of the breeding birds of Tenerife; it includes an excellent bibliography of publications on birds of the Canaries.

Pérez Padrón, F. 1986. THE BIRDS OF THE CANARY ISLANDS. Aula de Cultura, Cabildo Insular de Tenerife.
A recent translation of a Spanish text, illustrated with colour photographs. It includes a useful list of non-breeding birds.

237

Pizarro, M. 1984. *PECES DE FUERTEVENTURA*. Gobierno de Canarias.
A lavishly illustrated account of the fish of Fuerteventura, the majority of which also occur in the waters around Tenerife. In Spanish.

Rochford, N. 1984. LANDSCAPES OF TENERIFE - A COUNTRYSIDE GUIDE. Sunflower Books, London.
A useful pocket-sized book including details of 33 walks of varying length and difficulty. The author's interest lies primarily in the scenery. Available in Britain.

Santos, A. 1979. *ARBOLES DE CANARIAS*. Editorial Interinsular Canaria, Santa Cruz de Tenerife.
A well illustrated book concentrating on the endemic trees of the island; especially useful for identifying trees of the laurel forest. In Spanish.

General books referred to

Cramp, S. (editor). 1977-..... BIRDS OF THE WESTERN PALAEARCTIC. Oxford University Press.

Fischer, W. et al. (editors). 1981. SPECIES IDENTIFICATION SHEETS FOR FISHERY PURPOSES: EASTERN CENTRAL ATLANTIC, Vols.1-5. F.A.O.

Heath, J. (editor). 1979. THE MOTHS AND BUTTERFLIES OF GREAT BRITAIN AND IRELAND, Vol.19 . Curwen, London.

Heywood, V.H. (editor). 1985. FLOWERING PLANTS OF THE WORLD. Croom Helm, London.

Higgins, L.G. and Riley, N.D. 1983. A FIELD GUIDE TO THE BUTTERFLIES OF BRITAIN AND EUROPE. Collins, London.

Macdonald, D. (editor). 1984. ENCYCLOPAEDIA OF MAMMALS. George Allen and Unwin, London.

Maire, R. 1952-1980. *FLORE DE L'AFRIQUE DU NORD*, Vols.1-15. Lechevalier, Paris.

Polunin, O. and Huxley, A. 1981. FLOWERS OF THE MEDITERRANEAN. Chatto and Windus, London.

Vaurie, C. 1959-1965. THE BIRDS OF THE PALAEARCTIC FAUNA. H.F. and G. Witherby, London.

Voous, K.H. 1973 and 1977. LIST OF RECENT HOLARCTIC BIRD SPECIES. Ibis, Vol.117, p.612-638 and Vol.119, p.376-406.

Watson, L. 1981. WHALES OF THE WORLD. Hutchinson, London.

Whitehead, P.J. et al. (editors). 1984 and 1986. FISHES OF THE NORTH-EASTERN ATLANTIC AND THE MEDITERRANEAN. UNESCO.

Wunderlich, J. 1987. THE SPIDERS OF THE CANARY ISLANDS AND MADEIRA. Tropical Scientific Books, Langen, Germany.
This is an up-to-date technical book, mainly in German.

BUS INFORMATION FOR EXCURSIONS

The bus services on the island are excellent. The main bus company, *TITSA*, covers most of island except for the peninsulas of Anaga and Teno. Various small bus companies fill the gaps left by the TITSA services.

TITSA. Tel. 21 81 22. Principal bus station, Avenida Tres de Mayo, Santa Cruz de Tenerife, with subsidiary bus stations in other main towns. Many bus stops have lists of times of local services. A complete timetable can usually be obtained from the main bus station. There are frequent express buses between the larger towns, and many additional services to small communities.

TRANSPORTES DE SAN ANDRES. Tel. 28 12 11. Runs from Avenida Anaga in Santa Cruz de Tenerife to Taganana and Almáciga in Anaga, via the tunnel at El Bailadero. Frequent service to San Andrés, and more restricted service to Taganana and Almáciga.

TRANSMERSA. Tel. 25 07 40. Runs from La Laguna to various destinations in the mountains of Anaga.

TRANSVIMAR. Tel. 25 25 48. Frequent service from La Laguna to Punta del Hidalgo via Bajamar.

The bus times given here were all correct at the time of going to print. However, you would be well advised to check the times before setting out - especially on an excursion where the buses are infrequent.

EL MEDANO

"*TITSA*" bus 342 from *PUERTO DE LA CRUZ* to *EL MEDANO* 08.30 Mon, Thurs, Sat, Sun. Return from *EL MEDANO* 15.30. A long bus ride but a magnificent scenic route through *LAS CAÑADAS*. Also hourly "*TITSA*" bus 111 from *SANTA CRUZ* to *PLAYA DE LAS AMERICAS*; get off at the exit east of the airport, and walk.

MALPAIS DE GUIMAR

"*TITSA*" bus 120 from *SANTA CRUZ* to *PUERTO DE GUIMAR* every hour from 06.45-20.45 and 21.15; Sat, Sun and holidays every two hours from 06.45-20.45 and 21.45. Return bus leaves *GUIMAR* every hour from 06.15-19.15, and 20.20 and 21.15; Sat, Sun and holidays every two hours from 06.45-20.45.

VOLCAN DE GUIMAR

Hourly "*TITSA*" bus 111 from *SANTA CRUZ* to *PLAYA DE LAS AMERICAS*, and other buses along the *autopista*. Get off at the *GUIMAR* exit and walk back to the second fly-over bridge north of the exit. Return by any bus along the *autopista*.

TENO BEYOND BUENAVISTA

"TITSA" bus 363 from *PUERTO DE LA CRUZ* to *BUENAVISTA*. Every hour from 06.30-20.30, or bus 355 from *ICOD* at 06.15. Return from *BUENAVISTA* every hour from 05.30-19.30 (and 20.30, but change at *ICOD*). Takes 1 3/4 hrs. Take a taxi at *BUENAVISTA* (if you can find one) or walk or hitch.

BARRANCO DE IGUESTE

"TRANSPORTES DE SAN ANDRES" bus from *Avenida Anaga* in *SANTA CRUZ* to *IGUESTE*. Mon-Sat 06.00, 12.00, 14.30, 18.00; Sun and holidays 07.25, 08.00, 09.30, 12.00, 14.30, 18.00. Return from *IGUESTE* 06.00, 06.50, 08.00, 12.40, 16.00, 18.45, 20.10; Sun and holidays 08.30, 10.15, 13.30, 16.30, 18.00, 20.00. Check with bus company. Takes 30min.

LADERA DE GUIMAR

"TITSA" bus 125 from *SANTA CRUZ* to *GRANADILLA* via *GUIMAR* at 05.00, 09.30, 14.15, 17.30; get off at highest point of road 3.5km (2mi) beyond *GUIMAR*. Return bus leaves *GRANADILLA* 05.05, 09.15, 13.30, 18.30.

DEGOLLADA DE CHERFE

Bus from *ICOD DE LOS VINOS* to *GUIA DE ISORA* via *SANTIAGO DEL TEIDE*, about every two hours; then take taxi or walk. Return bus from *SANTIAGO DEL TEIDE* about every two hours. Time about 55min. Inquire about bus times at *ICOD*.

BARRANCO DEL INFIERNO

"TITSA" bus 471 from *GRANADILLA* to *ARMEÑIME* via *ADEJE, LOS CRISTIANOS* and *PLAYA DE LAS AMERICAS*. It leaves *PLAYA DE LAS AMERICAS* at 06.15 and then hourly at 25 minutes past the hour until 19.00. Return from *ADEJE* at about 20 minutes past the hour, hourly until 20.20 and also 20.40.

ANAGA BEYOND EL BAILADERO

"TRANSPORTES DE SAN ANDRES" bus from *Avenida Anaga* in *SANTA CRUZ* to *TAGANANA* and *ALMACIGA* via *EL BAILADERO*. Mon-Fri 06.50, 13.00, 17.00, 18.00, 19.00; Sat 07.00, 13.00, 17.00, 19.00; Sun 07.30, 13.00, 17.00, 19.00; holidays 08.30, 13.00, 17.00, 19.00. Ask for *el túnel*: this is the road tunnel that goes through the mountains (and under the ridge road) 17km (10.5mi) north of *SAN ANDRES*. The start of the tunnel is 1.5km (1mi) beyond a left turn signed to *"EL BAILADERO"*. Don't get off at that turning, or you would be involved in a long steep walk. Return from *el túnel* Mon-Fri 06.00, 08.05, 14.05, 18.05, 19.15, 20.15; Sat, Sun and

holidays 08.05, 14.05, 18.05, 20.15. Check times with bus company. Takes 35min.

PICO DEL INGLES

"TRANSMERSA" bus from *LA LAGUNA* to various destinations in *AN-AGA* via *PICO DEL INGLES* at 07.00, 09.00, 10.15, 13.15, 15.15, 16.00, 19.00 - but not every day. Check times with bus company.

MONTE DEL AGUA

Bus from *ICOD DE LOS VINOS* to *GUIA DE ISORA* via *ERJOS DEL TANQUE* about every two hours. Return from *ERJOS DEL TANQUE* about every two hours. Check bus times at *ICOD*. Takes about 40 min.

CHANAJIGA

The closest bus is *"TITSA"* 347 from *OROTAVA* to *CRUZ SANTA* via *BENIJOS* and *PALO BLANCO* at 06.00, 08.15, 10.15, 12.15, 15.15, 17.15, 19.15; ask for the *LLANADAS* turn-off and walk about 6.5km (4mi). Return bus leaves *CRUZ SANTA* at 06.30, 09.20, 11.20, 13.20, 16.20, 18.20, 20.20.

AGUAMANSA

"TITSA" bus 345 from *PUERTO DE LA CRUZ* to *LA CALDERA* (just beyond *AGUAMANSA*) every hour from 08.00-18.00; the bus goes only as far as *AGUAMANSA* at 06.45, 7.00, 19.00, 20.00. Return from *LA CALDERA* every hour from 09.30-21.30 (and also from *AGUAMANSA* at 06.00, 07.00, 07.45, 08.30).

LAS LAJAS

"TITSA" bus 342 from *PUERTO DE LA CRUZ* to *EL MEDANO* via *LAS LAJAS* 08.30 Mon, Thurs, Sat and Sun only. Return bus leaves *VILAFLOR* 16.45 on same days. Takes about 3hrs including several 10min stops. A magnificent scenic trip through *LAS CAÑADAS*.

EL LAGAR

Only possible as far as *LA GUANCHA*. *"TITSA"* bus 354 from *PUERTO* to *LA GUANCHA* every hour from 05.15-20.15. Return bus leaves *LA GUANCHA* every hour on the hour from 06.00-19.00, and at 20.15.

LA FORTALEZA

"TITSA" bus 348 (or 342) from *PUERTO DE LA CRUZ* to *LAS CAÑADAS* via the *Centro de Visitantes* (Visitor Centre) at 08.30 only. Return from Visitor Centre Tues, Wed, Fri 16.40 (approx.); Mon, Thur, Sat, Sun 18.10 (approx.). Takes 1hr 35min including 10min stop at *OROTAVA*. (From *LOS CRISTIANOS* you could take bus 342 on Mon, Thur, Sat or Sun in the afternoon and return on a later day.)

PICO DEL TEIDE

"TITSA" bus 348 (or 342) from *PUERTO DE LA CRUZ* to *LAS CAÑADAS* past the bottom of the *MONTAÑA BLANCA* track and *teleférico* terminal, at 08.30 only; ask for *MONTAÑA BLANCA* if you intend to walk up the mountain. Return from *teleférico* Tues, Wed, Fri 16.15 (approx.); Mon, Thur, Sat, Sun 17.45 (approx.). Takes 2hrs including 10min stop at *OROTAVA*.

ROQUES DE GARCIA

"TITSA" bus 348 (or 342) from *PUERTO DE LA CRUZ* to *LAS CAÑADAS* via the *Parador* (government-run hotel) at 08.30 only. Return from *Parador* Tues, Wed, Fri 16.00 (approx.); Mon, Thur, Sat, Sun 17.30 (approx.). Takes 2¼ hrs including a 10min stop at *OROTAVA*.

LAS NARICES DEL TEIDE

"TITSA" bus 342 from *PUERTO DE LA CRUZ* to *EL MEDANO*, via *BOCA DE TAUCE* in *LAS CAÑADAS*. 08.30 Mon, Thurs, Sat and Sun only. Get off at *BOCA DE TAUCE*, and walk 1km along road C823. Return bus leaves *VILAFLOR* 16.45 on same days. Takes about 2 1/2hrs, including several 10min stops.

CUEVA DE SAN MARCOS

"TITSA" Bus 354 from *PUERTO DE LA CRUZ* to *PLAYA DE SAN MARCOS* every hour from 05.15-20.15. Return bus leaves *PLAYA DE SAN MARCOS* every hour on the hour from 06.00-20.00.

PUNTA GOTERA

"TRANSVIMAR" bus from *LA LAGUNA* to *PUNTA DE HIDALGO* via *BAJAMAR* every half hour from 06.30-21.30. Ask for *El Club Náutico*. Return from *BAJAMAR* every half hour. Check times with bus company. Takes 35mins.

PUNTA DEL HIDALGO

"TRANSVIMAR" bus from *LA LAGUNA* to *PUNTA DEL HIDALGO* every half hour from 07.00-21.30. Return from *PUNTA DEL HIDALGO* every half hour. Check times with bus company. Takes 40mins.

FERRY TO *LA GOMERA*

There is a linking bus from *SANTA CRUZ* main bus station to the *"FERRY GOMERA"* terminal in *LOS CRISTIANOS* in the south of the island, leaving at 08.00 and 18.00. The *"Benchijigua"* leaves from *LOS CRISTIANOS* at 10.00 and 20.00 each day. It returns from *SAN SEBASTIAN* on *LA GOMERA* at 08.00 and 18.00.

INDEX

This index includes all plants and animals mentioned in the book, with the exception of the fish. Only the very few fish mentioned in Sections 1 and 2 are included (refer to the Fish pages of Section 3). **Page numbers are given for references to Sections 1 and 3, while excursion numbers (eg. Exc.4,5) are given for Section 2.**